I0469517

STRUCTURAL UNIFICATION

OF

QUANTUM MECHANICS

AND

RELATIVITY

Volume I

Emile Grgin

AuthorHouse™
1663 Liberty Drive, Suite 200
Bloomington, IN 47403
www.authorhouse.com
Phone: 1-800-839-8640

© 2007 Emile Grgin. All rights reserved.

No part of this book may be reproduced, stored in a retrieval system, or transmitted by any means without the written permission of the author.

First published by AuthorHouse 12/6/2007

ISBN: 978-1-4343-1048-4 (sc)

Library of Congress Control Number: 2007902896

Printed in the United States of America
Bloomington, Indiana

This book is printed on acid-free paper.

authorHOUSE®

Preface

In the title of this book, "structural unification" refers to a merging of two physical theories into a *single mathematical structure*. We impose the following conditions on such a merging:

Symmetry: The component theories must enter unification on the same footing. This is to eliminate approaches to unification that treat as more fundamental the theory which happens to be better known.

Completeness: Within its domain of applicability, the unification must yield no unphysical theorems. This is to eliminate unifications that require *ad hoc* adjustments when faced with the real world.

Irreducibility: The unification must yield more physics than can be derived from the given theories taken independently. This is to eliminate from consideration physically empty unifications.

Maxwell's electromagnetic theory is an example of structural unification: In covariant form, this theory is *symmetric* because the electromagnetic tensor subsumes the electric and magnetic fields with equal weights; it is *complete* because none of its theorems has to be rejected as unphysical for clashing with observations; it is *irreducible* because it yields at least electromagnetic radiation as new physics.

In contrast, quantum field theory is not a structural unification of quantum mechanics and relativity for not being a "single mathematical structure". It is not symmetric either: Canonical quantization grafts quantum properties on relativistic fields, thus treating these properties as after-thoughts. This cannot possibly be the way of nature.

We reconsider in this book an approach to the generalization of quantum mechanics based on the substitution of a structurally richer number system for the field of complex numbers.

This is a very old idea, but it has been tested only under the assumption that the algebraic norm in the new number system must be a non-negative real number (an "algebraic norm" is an object like $\|z\|^2 = z^*z$). This limits the search for the unknown number system to the division algebras. Among these algebras, the only acceptable candidate is the field of quaternions — but this field does not support a structural unification of quantum mechanics and relativity.

Our approach is more general in that the values of the algebraic norm function are not subject to any *a priori* conditions. This type of generalization has a precedent: Relativistic spacetime came into being when Minkowski broke the historical bondage of geometry to the non-negative metric norm that goes back to Pythagoras.

As shown in previous publications by the author, a unifying number system does exist, but not as a structure already investigated within mathematics. We refer to it as the "algebra of quantions". It has all the properties necessary to supplant the field of complex numbers in quantum mechanics, but none that have to be discarded as unphysical. Moreover, *quantions are relativistic,* so that the generalized theory is "inherently relativistic". The unification is thus structural without any effort on our part to make it so. The existence of a locally Minkowskian spacetime, as well as the Schrödinger and Dirac equations, *do not have to be postulated:* They are quantionic theorems.

The following parallel statements put the quantionic approach to physics in a historical perspective.

Observation: The mathematical obstruction which made it impossible to extend classical mechanics to quantum phenomena during the first quarter of the 20th century was the number system of classical physics, this system being the field of real numbers.

Assertion: The mathematical obstruction which made it impossible to extend quantum mechanics to relativistic phenomena during the last three-quarters of a century is the number system of quantum physics, this system being the field of complex numbers.

It has been recognized since then that a nonrelativistic quantum mechanics can be built only over the structurally much richer field of complex numbers.

It is the contention of the present work that a relativistic quantum mechanics can be built only over the structurally much richer algebra of quantions.

The basis for the structural unification is an algebraic merging of the complex numbers with the linear Minkowski space. If this unification were completed — and it seems reasonable to expect that it can be — it could be characterized by paraphrasing Minkowski:[1] *Relativistic spacetime by itself and the field of complex numbers by itself are to sink into oblivion, and only a kind of union of these concepts is to represent reality.* The field of complex numbers is meant here only in its role of number system for relativistic quantum physics.

The present work is organized around four topics:
— The first chapter is dedicated to an introductory discussion of number systems, both in general and from the point of view of physics.
— Part I develops the mathematical theory of quantions *ab initio*.
— Part II is about the physical applications of quantions that have been developed to date.
— The annotated bibliography provides an overview of the development of the quantionic approach to unification over the last thirty years. Since the present book is self-contained, these references are not prerequisite reading.

The presentation of the quantionic approach to physics, as developed in this book, has greatly profited from many discussions with Nikola Zovko, from the Institute Ruđer Bošković in Zagreb; with Guido Sandri and Anton Mavretič, both at Boston University; and with Florin Moldoveanu, from the National Institute of Physics in Bucharest. Zovko's contribution was also essential: At an early stage of the development of the quantionic algebra, he suggested a generalization of Born's interpretation that plays a key role in the derivation of the quantionic equations of motion. I call it "Zovko's interpretation".

New York, November 2007.

---

[1] In 1908, at a lecture in Köln, Minkowski characterized the space that now bears his name by a very quotable succinct declaration. In free translation, his statement reads: "From this day on, space by itself and time by itself are to sink into oblivion, and only a kind of union of these concepts is to represent reality."

Contents

Chapter 1

Number Systems

Only two types of numbers support all of contemporary physics: Classical mechanics and relativity are built over the real numbers; nonrelativistic quantum mechanics and quantum field theory are built over the complex numbers. Since both types of numbers are well known and taken for granted by students, the general concept of a number system and the role number systems play in physics are never discussed in physics textbooks.

The essential point of the present work is that it introduces a new number system for relativistic quantum physics. This system, referred to as the "algebra of quantions", is also new as a mathematical structure. It thus seems appropriate to dedicate this introductory chapter to a discussion of number systems — first from the point of view of mathematics (Section 1.1), and then of physics (Section 1.2).

1.1 Number systems in mathematics

Every number system is an algebra (meaning a linear space equipped with a product) but very few algebras are number systems. Yet, there are no formal criteria that tell us whether a particular algebra ought to be regarded as a number system. This was clearly stated by the mathematician and mathematical logician Abraham Robinson, who made some very enlightening observations on the subject in his philosophical papers. The two quotations that follow are from his Brower Memorial

Lecture, titled "Standard and Nonstandard Number Systems".[1]

Robinson stated the justification for referring to some mathematical structure as a number system as follows:

> It is, inevitably, a matter of convention, which algebraic or arithmetical structures are to be honored by the name of *number systems*. Generally speaking, the more naturally a new system seems to grow out of a *number system* of acknowledged standing, and the more similar to such a system its features, the more likely it is that the new entities also will be regarded as numbers.

We shall see that the algebra of quantions satisfies these criteria. It may thus be regarded as a number system from the point of view of mathematics.

The next quotation sheds light on the motivations for introducing new number systems:

> The invention or discovery of new number systems should not be regarded as an end in itself. In fact, each of the two developments that I have described occurred in response to a natural demand. We may ask whether there still exists in mathematics a situation which may give rise to a system of numbers as yet unknown. The answer is that such a situation not only exists but that it is close to some of the most exciting problems of contemporary mathematics.[2]

The algebra of quantions was developed by the author in response to a 'natural demand' in the foundations of physics: The necessity of unifying quantum mechanics with general relativity. While this motivation stems from physics, the problem itself is mathematical.

The class of all mathematical structures that may be viewed as number systems is open ended, but very few are well known. The

---

[1] Read in Leiden, April 26, 1973. Reprinted in *"Selected Papers of Abraham Robinson" Vol. 2: Nonstandard Analysis and Philosophy*. Editors: Keisler, Körner, Luxemburg, and Young. Yale University Press. New Haven. 1979. (Page 426.)

[2] The developments in question are nonstandard numbers and nonstandard analysis.

mutual relationships between the better-known systems (with the addition of the algebra of quantions) are illustrated in Figure 1.1. Those of relevance in physics are distinguished by bold cartouches.

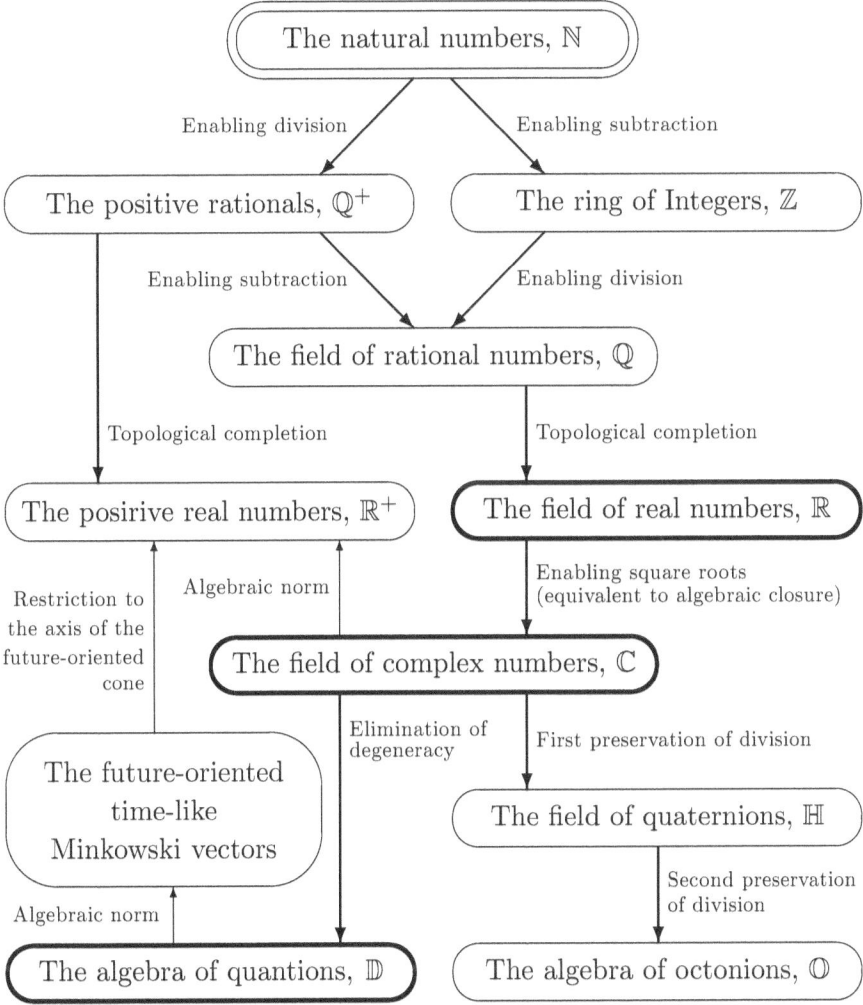

Figure 1.1. The major number systems.

The down-pointing arrows represent what Robinson informally referred to as "growing out of". The technical term for this procedure is "extension". For the inverse procedure, it is "restriction". Thus, as an example of unambiguous terminology, we may say that "the field $\mathbb{R}$ of real numbers is the topologically complete extension of the field $\mathbb{Q}$ of rational numbers". Conversely, "the field $\mathbb{Q}$ is a restriction of the field $\mathbb{R}$". There are infinitely many other restrictions of $\mathbb{R}$, but $\mathbb{Q}$ is distinguished as the 'minimal subfield of characteristic zero'.

To take a walk down the graph of number systems, let us begin at the system of natural numbers, $\mathbb{N} = \{1, 2, 3, \cdots\}$, whose informal origins are lost in prehistory. Borrowing Robinson's intuitive terminology, we take $\mathbb{N}$ to be a number system of "acknowledged standing". The arithmetic operations it admits without restrictions are addition, multiplication, and exponentiation.

The theory of proportions developed by Eudoxus in 360 BC eliminated the exception related to division by extending the system $\mathbb{N}$ to the system $\mathbb{Q}^+$ of all fractions p/q, where p and q are relatively prime numbers. This theory was formulated in the language of geometry, which was the only one trusted in Classical Greece, but its content is essentially the same in the modern algebraic formalism. This procedure is indicated in the graph as 'enabling division'. Thus, the system $\mathbb{Q}^+$ "grew naturally" out of the acknowledged system $\mathbb{N}$. It is also "very similar" to $\mathbb{N}$ for admitting the same operations — though division is no longer subject to exceptions. Clearly, the system $\mathbb{Q}^+$ of positive rational numbers may be "honored by the name of number system".

We should point out that "enabling" is not standard terminology in mathematics. It is meant here to convey the meaning of *making possible without exceptions an operation that is not universally possible, but is well defined whenever it is possible.*

The other two exceptions were similarly eliminated, but only in the Late Renaissance. They are indicated on the graph by the self-evident labels "enabling subtraction" and "enabling square roots". The latter implies, though not obviously, that all rational exponents are allowed. Historically, the extension to complex numbers took place while the definition of the real numbers was still geometric, that is, long before they were put on a rigorous axiomatic foundation by Dedekind.[3] The complex numbers were forced upon the mathematical scene in 1545 by

---

[3] Drawing a historically faithful graph would require a long introduction to algebraic number systems, which, unlike those of Figure 1.1, are neither widely known nor relevant in the sequel.

Cardano, but it was only in 1833 that Hamilton introduced a reinterpretation of these numbers which eliminated the ontological objections to the imaginary unit and opened the door to generalizations.

The most remarkable property of the field of complex numbers is its algebraic closure. This means that no further extensions are possible by the process of "elimination of exceptions". The reason is that no exceptions are left in the performance of arithmetic operations. This theorem, known as "the fundamental theorem of algebra", was first proved by Gauss in his doctoral dissertation of 1799. He then proved it thrice again over the following fifty years. The modern proof, which is much simpler, is based on topological arguments.

The extension procedure that led to the field of real numbers is topological completion. In the fifth century BC, the Pythagoreans discovered — by a proof still viewed as one of the most beautiful in all of mathematics — that the diagonal of a square is not commensurate to its side, which implies that $\sqrt{2}$ is not a rational number. The confusion this discovery produced in mathematics was clarified by Dedekind only at the end of the 19th century. The reason it took so long is that the solution is not of an exclusively algebraic nature: It relies on the topological concept of continuity, which was identified only in the 19th century. In modern parlance, real numbers are "equivalence classes of Cauchy sequences of rational numbers". They may seem intuitively elementary, but their formal definition is not.

With the introduction of the complex numbers, the existence of all inverse objects was guaranteed. This includes the roots of polynomials with arbitrary complex coefficients. Consequently, further extensions of the field of complex numbers could be based only on new ideas.

Thus, the extensions to quaternions and octonions were motivated by mathematical curiosity, not by mathematical necessity. They are based on the relaxing of some axiomatic properties of the complex numbers. Since some desirable properties might be lost in the process, it is necessary to specify beforehand which properties are to be preserved.

Preserving the possibility of division by all non-vanishing elements has been considered most important because it leads to interesting theorems. This requirement is essentially equivalent to the condition that the norm of type z^*z be non-negative.

The number systems that admit unrestricted division (except by zero) are called "division algebras". Thus, real and complex numbers are evidently division algebras. In 1843 Hamilton discovered the quaternions by dropping commutativity, and in 1845 Cayley discovered the octonions by dropping associativity as well. Finally, in 1895, Hurwitz proved that the family of division algebras consists exactly of these four number systems.

The extension of the complex numbers to quantions was initially motivated by physical considerations, but it may also be viewed — and more fruitfully so — as the "elimination of a degeneracy" of the complex numbers (see page 17).

Let us now turn to the meaning of the three up-pointing arrows. The set $\mathbb{R}^+$ is the image of $\mathbb{C}$ under the algebraic norm function $z \mapsto z^*z$. On the other hand, the corresponding algebraic norm of quantions, introduced in Part I by the same formal function, is not a positive real number but a future oriented Minkowski vector. Since the complex numbers are special quantions, the respective algebraic norms are also mutually related — both geometrically and physically:

(a) The set $\mathbb{R}^+$ of non-negative real numbers is the axis of the future-oriented null cone.

(b) Born's interpretation of the algebraic norm $\psi^*\psi \in \mathbb{R}^+$ of complex numbers is a special case of Zovko's interpretation of the algebraic norm of quantions.

These two points characterize the link between standard quantum mechanics and its quantionic generalization.

Other major number systems not shown in Figure 1.1 are:

(1) The integers modulo a prime number (Gauss 1801).

(2) The various types of algebraic numbers (mid 19th century).

(3) The p-adic numbers (Hensel, 1897).

(4) The nonstandard numbers (Robinson, 1966).

(5) The Grassmann algebras (Grassmann, 1844).

(6) The Clifford algebras (Clifford, 1882).

The first four, rooted in arithmetic, are universally recognized as number systems. The last two, rooted in geometry, are regarded by some authors as candidate number systems for physics.

1.2 Number systems in Physics

In mathematics, the question whether a given algebraic structure ought to be viewed as a number system is adequately answered by the first quotation on page 2. In physics, the question was never considered. We suggest the following criterion:

An algebra may be viewed as a number system if physical states are represented by the elements of the algebra in question.

In classical mechanics, states and observables are both real: States are represented by probability distributions in phase space, observables by real functions in the same space. The number system of classical mechanics is thus the field of real numbers. We don't speak of states in relativity, but both special and general relativity are formulated over the real numbers. Both are classical, in the sense of 'non-quantum'.

In quantum mechanics, observables preserve to some extent their intuitive classical meaning. They are 'measured' by real numbers, in the sense that their eigenvalues are real. The states, however, are represented by complex numbers. The number system of quantum mechanics is thus the field of complex numbers.

The following three questions have to be answered before we can undertake a search for a structurally unifying number system:

First question: Do the complex numbers have any properties that need not (or must not) be preserved in the new number system?

Second question: Which properties of the complex numbers must be preserved in the new number system?

Third question: Are there any properties that must be present in the new number system because they are needed for structural unification, but are not present in the complex numbers?

The answers are collected at the bottom of page 19.

Discussion of unification

If we think of quantum mechanics and relativity not as physics but as axiomatically defined mathematical theories, the unification problem is not a problem in physics but in mathematics. It may be stated as follows:

Find the simplest mathematical structure which contains the two theories as substructures.

The idea is illustrated in the following diagram, where 'relativity' is left unspecified but is ultimately meant to be general relativity. This point is revisited on page 12.

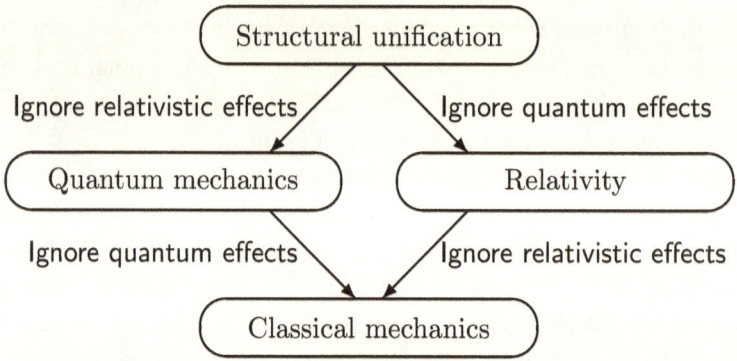

The first step in the search for a structural unification consists in selecting a search strategy. Most approaches that have been considered in the past fall into two categories: geometric and algebraic.

The *geometric approaches* consist in generalizing the relativistic structure of spacetime while initially ignoring quantum mechanics. The latter is expected to come out of geometry as a theorem. An example of this strategy is Penrose's theory of twistors.

The *algebraic approaches* consist in generalizing the algebraic structure of quantum mechanics while initially ignoring relativity. The latter is expected to come out of algebra as a theorem. If one begins with the Hilbert space formulation of quantum mechanics, two possibilities suggest themselves:

One may try to modify the axioms of Hilbert space while preserving the field of complex numbers as the underlying number system. While modifying the axioms of a theory is a very common type of generalization in the creation of new mathematics, no acceptable modification of the axioms of Hilbert space could ever be found. It is now believed that none exists. For this reason, one usually says that Hilbert space quantum mechanics is "rigid".

Alternatively, one may try to substitute a structurally richer number system for the field of complex numbers while preserving the axioms of Hilbert space. (It is true that a new number system would imply some modifications of Hilbert space, but these are not to be confused with outright modifications of the axioms.) This idea has been tested more than half-a-century ago by assuming that the new number system is the field of quaternions, but quaternionic quantum mechanics proved to be as nonrelativistic as the standard theory built over the field of complex numbers.

These negative outcomes led to the nearly universally accepted view that a unification of the type we refer to as "structural" does not exist. This proved to be an overhasty conclusion because the non-existence of a unifying number system that could supplant the field of complex numbers has never been demonstrated. What was shown is that the quaternions are not the solution — but the quaternions were merely the first and only guess at a solution.

The following two observations should help us understand what led to the logical error in question.

First observation: In the author's opinion, it was unthinkable that the unknown of a physical problem could be a number system that generalizes the field of complex numbers in a somehow minimal way, but is nevertheless still unknown within mathematics. The tacitly held conviction was that all reasonable generalizations of the complex numbers are the division algebra. Within this belief, the minimal modification of the field of complex numbers is the field of quaternions. There was also a physical argument that favored division algebras: It was taken for granted that Born's interpretation of the function $\rho = \psi^*\psi$ as a probability density must be formally preserved in the transition to a new number system. This implied that the latter must have a non-negative norm, which further implied that it must be a division algebra.

Second observation: Wanting to preserve Born's interpretation is natural if the objective is to generalize nonrelativistic quantum mechanics out of curiosity. It is self-defeating if the objective is to explore the possibility of structurally merging quantum mechanics with relativity. To see why, it suffices to realize that the density ρ is not a scalar in Dirac's relativistic theory of fermions. It is the time component of a four-current. This observation suggests a general insight:

No division algebra can support a structural unification of quantum mechanics and relativity.

It follows that if a unifying number system exists, it can be discovered only as a new algebraic structure, that is, as an algebra that must be constructed *ab initio* from some general characteristics of the unification problem itself. It would be futile to search for the solution in a mental catalogue of mathematical structures that have already been investigated as number systems within mathematics.

The contention of this book is that a unifying number system does exist, and that it is the algebra of quantions formally developed in Part I. The genesis of this algebra is not elevant to the objectives of the present work. It is also too long to be reproduced on these pages. For our purposes, it suffices to make the quantions heuristically plausible. This is done in the remainder of this chapter.

The annotated bibliography ought to be helpful to readers who might be interested in the sequence of ideas that led to quantions.

Construction of the algebra of quantions

The diagram on page 8 characterizes structural unification from the viewpoint of physics. To redraw it from the viewpoint of mathematics, one considers only the underlying mathematical structures:

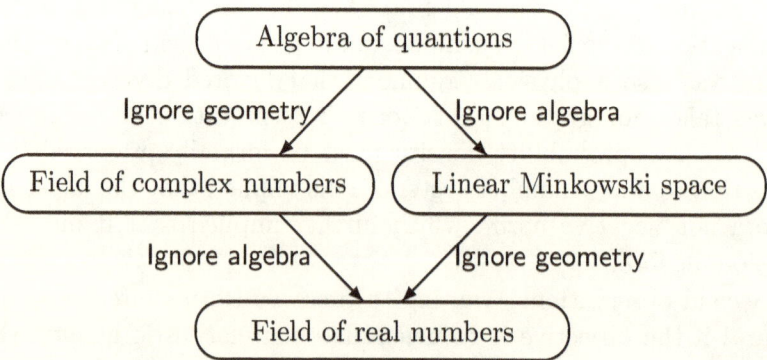

In this diagram, "Algebra of quantions" stands for the underlying number system of a hypothetical "Structural unification". The underlying structures of quantum mechanics, of relativity, and of classical mechanics, are, respectively, the field of complex numbers, the linear Minkowski space, and the field of real numbers.

It is evident that the smallest structure which contain both the complex numbers and the real Minkowski vectors is the complex linear Minkowski space. The previous diagram may thus be rewritten in terms of the components of a complex Minkowski vector:

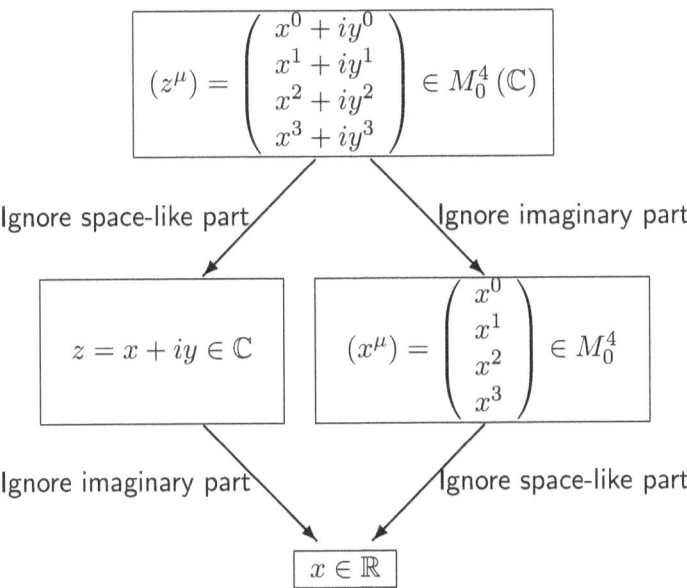

However suggestive, this diagram represents only an intermediate step in the construction of a unifying algebra. The reason is that three properties are still missing. Let's refer to them as *algebra, derivation,* and *curvature.* What follows is a brief outline of how these properties can be introduced.

Algebra

The complex linear Minkowski space $M_0^4 (\mathbb{C})$ is only a linear metric space. In order to upgrade it to an algebra, one must find a bilinear function that assigns a single complex Minkowski vector to every pair of such vectors. Clearly, the function in question (called a product) may not be selected arbitrarily because the algebra must be 'unifying'. This means that it must naturally yield the characteristic objects of relativity and quantum theory. These objects are the Minkowski norm and a physically meaningful generalization of the norm $\psi^*\psi$ of complex numbers.

The minimal algebra which satisfies these requirements is the full algebra of 2×2 matrices

$$q = \begin{pmatrix} q_1 & q_3 \\ q_2 & q_4 \end{pmatrix},$$

where the q_i are arbitrary complex numbers. The complex Minkowski vectors and the complex 2×2 matrices are formally related by the Pauli matrices.

The Minkowski norm appears as the determinant $\det(q)$, while the norm $\psi^*\psi$ generalizes to $q^\dagger q$, which is a Hermitian matrix. As expected (see the remark at the top of page 10), this object is not a positive real number. It's physical meaning is given by Zovko's interpretation.

Derivation

By itself, the algebra of complex 2×2 matrices does not solve the unification problem because it subsumes only the linear Minkowski space (whose invariance group is the Lorentz group) while differential field equations are defined in an affine Minkowski space (whose invariance group is the Poincaré group). It follows that we are to find a structurally distinguished way of introducing the translation group, or, equivalently, a derivation operator. This means that we have to introduce new mathematical degrees of freedom without losing the algebra of complex 2×2 matrices in the process.

A unique solution does exist (the uniqueness is not obvious): It consists in representing the algebra of complex 2×2 matrices by a subalgebra of the algebra of complex 4×4 matrices. The 16 matrix elements of the latter are thus linear combinations of the four complex variables q_1 to q_4. If the complex coefficients defining the linear combinations are constants, a derivation operator without affine connections is uniquely defined (Chapter 13) and the Schrödinger and Dirac equations follow as quantionic theorems — which is essentially the subject matter of Part II.

This is how far the quantionic approach to quantum-relativistic physics has been developed to date.

Curvature

From the point of view of differential geometry, an affine Minkowski space is a flat Riemannian space whose metric is locally Minkowskian. It follows that if the derivation operator mentioned above is to support general relativity, it must be modified by an affine connection that varies from point to point.

While this idea remains to be formally developed, it seems likely that the algebra of complex 4×4 matrices will not have to be extended to a larger structure. An indication that this might be so is that the 4×4 representation of the 2×2 algebra is not unique: There exists a multiple infinity of such representations, so that a sufficient number of mathematical degrees of freedom remains available to ensure consistency with the ten arbitrary spacetime functions that define the metric tensor in a four-dimensional Riemannian space.

The mathematical properties of quantions

By the end of Part I, all properties of individual quantions and of the algebra of quantions as a structure will have been formally derived. Yet, we can already anticipate the most important of these properties from the construction outlined above and from the heuristically justified physical requirements discussed below. The advantage of doing so at this time is that the concepts and theorems of Part I will appear better motivated as they are being introduced.

It is most convenient and most instructive to discuss the expected properties of quantions by comparing them to the properties of division algebras. This is done in Table 1.1 on the next page.

In this table, the characteristics of the algebra of quantions are doubly-framed for emphasis in the vertical rectangle. By comparing the entries "yes" and "no", one observes that the algebra of quantions is structurally intermediate between the field of complex numbers and the field of quaternions, even though it has twice as many dimensions as the latter. Thus, if one does not insist on division algebras, the minimal modification of the field of complex numbers is the algebra of quantions, not the field of quaternions.

The properties that are doubly-framed in the horizontal rectangle are those of relevance to both nonrelativistic quantum mechanics and to its quantionic generalization.

Algebras ⟍ Properties	Real Numbers $\mathbb{R}$	Complex Numbers $\mathbb{C}$	Quantions $\mathbb{D}$	Quater- nions $\mathbb{H}$	Octonions $\mathbb{O}$
Dimensions — real	$1 = 1+0$	$2 = 1+1$	$8 = 4+4$	$4 = 1+3$	$8 = 1+7$
complex	-	1	4	-	-
Commuta- tivity	yes	yes	no	no	no
Division structure	yes	yes	no	yes	yes
Associativity	yes	yes	yes	yes	no
Complex structure	no	yes	yes	no	no
Polar structure	no	yes	yes	no	no
Derivation structure	yes	yes	yes	no	no
Non degeneracy	no	no	yes	no	no

Table 1.1: Classification of number systems.

Let us discuss in turn each row in Table 1.1.

Dimension: This refers to the number of dimensions of the underlying linear space of the number system in question. The total number of basis vectors that span this space is written in the form $n = r + s$, where r is the number of real unit vectors and s the number of imaginary unit vectors. Thus, the field of real numbers has just one real dimension, while the algebra of octonions has one real and seven imaginary dimensions. If $r = s$, the underlying linear space may be viewed as a space of r complex dimensions.

Commutativity. Unlike the field of complex numbers it generalizes, the algebra of quantions is not commutative. Non-commutativity is often the price one has to pay for greater structural richness. One might even say that commutativity is somewhat 'uninteresting'. This observation is trite in our days, but it was not in 1843, when Hamilton discovered non-commutativity in a flash of insight that led to quaternions.

In the transition from classical to quantum mechanics, noncommutativity enters at the level of observables.

In the transition from nonrelativistic to inherently relativistic quantum mechanics — which is a synonym for structural unification — noncommutativity enters at the deepest mathematical level: in the underlying number system.

Division structure. This non-standard terminology refers to the possibility of division within an algebra, which is equivalent to saying that every element other than zero has an inverse. It is a mathematical theorem that such a division is possible if and only if the algebraic norm is a non-negative real number. This observation suggest that we should extract the physical meaning of the division structure both from the point of view of the inverse function and from the point of view of the norm. The reason is that the two approaches might yield different insights, even though they are mathematically equivalent.

The inverse: In the number system of nonrelativistic quantum mechanics (the field $\mathbb{C}$), the inverse of every element other than zero exists, but it plays no physical role whatsoever. Thus, the inverse ψ^{-1} of the wave function exists mathematically where $\psi \neq 0$, but it is conceptually meaningless because one never divides by a probability. It is true that the inverse of a determinant is needed to compute the inverse of a linear transformation, but since the only transformation groups relevant in quantum mechanics are unitary and unimodular (that is, of unit determinant), the number system of a generalized quantum theory *need not be* a division algebra.

The norm: As already concluded in the emphasized observation on top of page 10, the number system of a generalized quantum theory *must not be* a division algebra.

Associativity. The associativity of a number system guarantees the continuity of the group of unitary transformations built over it.

Thus, the unitary group over the real numbers is the orthogonal group; the unitary group over the complex numbers is the unitary group in its usual narrow sense; the unitary group over the quaternions is the symplectic group. These groups are continuous, which makes them candidates for the formulation of differential equations of motion. Thus, the physical meaning of associativity manifests itself in the smoothness of the time development of states (at least until a measurement 'collapses the wave function').

In contrast, the octonions are not associative, so that the unitary group over these objects is not continuous. It consists of the five sporadic groups in Cartan's classification. Thus, while a formal quaternionic quantum mechanics has been thoroughly developed,[4] there is no 'octonionic quantum mechanics'.

The associativity of the algebra of quantions guarantees the continuity of the "quantionic unitary group". We denote this group by $U_q(n)$. The quantionic unitary group acts in a quantionic Hilbert space which differs minimally from the ordinary Hilbert space defined over the complex numbers (Chapter 12).

Complex structure.[5] An algebra in which a complex conjugation is defined is said to have a complex structure if the two eigenspaces of the conjugation operator are isomorphic. An equivalent definition is that the underlying linear space has the same number of real and imaginary dimensions.

This property propagates to matrices, so that the linear space of all $n \times n$ matrices splits into two isomorphic subspaces of Hermitian and antihermitian matrices respectively.

In the language of physics, this property is known as the equivalence of observables and generators. It holds in standard and quantionic quantum mechanics, but not in the quaternionic version.

Polar structure. By "polar structure", we refer to the possibility of factorizing the elements of a number system into a "modulus" and a "phase factor". For complex numbers, the factorization is $z = re^{i\varphi}$.

---

[4] A modern monograph on the subject is: Adler, Stephen L. *Quaternionic Quantum Mechanics and Quantum Fields.* Oxford University Press, Oxford (1995).

[5] This terminology was introduced by Andrżej Trautmann in "On Complex Structures in Physics". In *Einstein's Path.* Essays in honor of Engelbert Schücking. A Harvey, Ed. Springer, New York (1999).

A similar expression exists for quantions (Chapter 10). The complex structure discussed above implies that the modulus and the phase factor are defined by the same number of parameters: one for complex numbers, four for quantions.

The physical meaning of the polar factorization manifests itself in the separation of observable and unobservable mathematical objects.

The observable part is the modulus. The interpretation of the square of the modulus, referred to as the algebraic norm, is probabilistic (Born's interpretation and its extension to Zovko's interpretation, Chapter 14).

The unobservable part is the phase factor. The totality of phase factors forms a gauge group which governs the classification of potentials in the field equations (Chapter 18).

Derivation structure.[6] We say that a number system has a derivation structure if a derivation operator compatible with the system's algebraic structure exists. The physical meaning of this property manifests itself in the existence of differential equations of motion, or field equations.

In the real and complex variables, x and z respectively, the definitions of the derivation operators $\frac{d}{dx}$ and $\frac{d}{dz}$ are intuitively evident, but in the algebra of quantions, intuition provides no guidelines. Even simple formal analogies break down due to noncommutativity. For these reasons, the construction of the quantionic derivation operator is a rather delicate matter. The solution developed in Chapter 13 is for differential field equations in the affine Minkowski space. No attempt is made in this book to extend the solution to Riemannian space.

Non-degeneracy. As shown in Table 1.1, this property belongs exclusively to quantions. It may thus be viewed as the key to structural unification — which justifies the following detailed discussion.

The word "degeneracy" usually refers to a structure-reducing coincidence. In the case of complex numbers, the coincidence is in the numerical values of two conceptually different functions: the "algebraic norm function" and the "metric norm function".

It is essential to view these functions as *conceptually different*. This may initially take some mental effort because a deeply-rooted prejudice

---

[6] The concepts of a "derivation structure" already exist in differential geometry, but in an area sufficiently distant from the present work to preclude confusion.

must first be overcome. By this prejudice, we erroneously take it for granted that the two functions are identical.

The algebraic norm: Let z be an arbitrary complex number viewed strictly as an element in the field of complex numbers — not as a vector in the complex plane. The inverse of z is $z^{-1} \equiv \frac{z^*}{z^*z} = \frac{z^*}{A(z)}$, where $A(z) \overset{def}{=} z^*z$ is a non-negative real number. Thus, the function $A(z)$ is a naturally distinguished norm function. Since its definition involves only algebraic concepts (complex conjugation and the product of complex numbers), we call it "the algebraic norm". In components:

$$A(z) = z^*z = (x+iy)^* (x+iy) = (x-iy)(x+iy) = x^2 + y^2.$$

The metric norm: Let us now ignore the field properties of complex numbers and interpret $z = x+iy$ as a vector (x,y) in the complex plane. Its length is $\|z\| = \sqrt{M(z)}$, where $M(z) \overset{def}{=} x^2 + y^2$ is a non-negative real number. Thus, the function $M(z)$ is a naturally distinguished norm function. Since its definition involves only a geometric concept (the Pythagorean metric), we call it "the metric norm".

The functions $A(z)$ and $M(z)$ are conceptually different. Hence, the fact that they happen to coincide,

$$A(z) = x^2 + y^2 = M(z), \tag{1.1}$$

is to be interpreted as a degeneracy. This observation is not merely a subtle philosophical point: *It is at the core of structural unification.* The reason is that viewing the equalities (1.1) as a degeneracy of the field of complex numbers inescapably suggest the following question: Does this field admit a non-degenerate algebraic extension?

This question was answered in [10]: Such an extension does exist, and it is unique. It is the algebra of quantions.

The simultaneous presence of the two norms in the same number system is the algebraic basis of structural unification:

Algebraic norm A	$\leftrightarrow$	Quantum mechanics
Metric norm M	$\leftrightarrow$	Relativity

Since the two norms are on the same footing, the symmetry of the structural unification is guaranteed — where "symmetry" is meant in the sense introduced in the Preface as the first requirement on page v.

In the author's experience, many physicists are not convined by the arguments presented above in favor of regarding the functions $A(z)$ and $M(z)$ as conceptually different. The problem is due to 'over-familiarity' with the equality (1.1). It might thus be helpful to point out that this equality is relatively recent, since the representation of a complex number $z = x + iy$ by a point (x, y) in a Euclidian plane (referred to as the "complex plane", or "Gaussian plane") became well known only in the mid-nineteenth century. This representation is obvious to us, but its chronology suggests that it entered mathematics as a daring idea: In October 1797, Gauss discovered the representation in question, but refrained from publishing it. Wessel discovered it independently in the same year and published it in 1798, but his work was ignored. Argand also discovered it independently and published it in 1806, but his work was barely noticed and soon forgotten. Finally, Gauss published it in 1831, introducing in the same paper the terminology "complex number". It is only from this point on that the idea spread throughout the mathematical community. Remarkably, this was 286 years after the complex numbers appeared on the mathematical scene.

We shall conclude this chapter with the answers to the questions raised on page 7. These answers mirror the organization of Table 1.1 on page 14. They also support the claim that the algebra of quantions is the number system naturally adapted to structural unification. We may even say *rigidly* so, for it admits of no modifications.

Answer to the first question: The properties that *need not be* preserved are commutativity and the division structure. They *are not* preserved in the algebra of quantions. Thus, this algebra does not carry any unnecessary (or even harmful) structural baggage.

Answer to the second question: The properties that *must be* preserved are associativity, complex structure, polar structure, and derivation structure. They *are* preserved in the algebra of quantions.

Answer to the third question: There is exactly one essential property not present in the division algebras: non-degeneracy. It is present exclusively in quantions.

Part I

THE MATHEMATICS OF QUANTIONS

Chapter 2

The Left Algebra of Quantions

Briefly stated, the present work is essentially the theory of the complex 4×4 matrices of Definition 1 on page 27 and of the concepts naturally related to them.

Since these block-diagonal matrices, referred to as quantions, are reducible to 2×2 matrices, it may seem less than believable that their algebra could play a role sufficiently fundamental "to be honored by the name of number system". It will become clear in the sequel that the reducibility is less important than it may seem because all sixteen elements of the 4×4 matrices have physical interpretations. It is just that a reducible subalgebra of these matrices plays a distinguished, purely algebraic role.

The following preview of the mathematical ideas that play key roles in the quantionic approach is meant to dispel the impression that Definition 1 is somehow arbitrary.

On page 12, the algebra of quantions was introduced as an isomorphic embedding of the algebra of complex 2×2 matrices in the algebra of complex 4×4 matrices. The 16 matrix elements of the 4×4 matrices are thus homogeneous linear functions of only four complex variables. The stability condition of an algebra imposes strong conditions on the coefficients that define the linear embedding functions. This is illustrated by the following concrete numerical example.

Let $\mathcal{A}$ denote the set of all matrices of the type

$$Q = \begin{pmatrix} 4y + z & 4z - x - 8y - 4w & 3x & -12y \\ -y & w + 2y & 0 & 3y \\ 0 & 2y & w & y \\ 2y & 2z - 4y - 2w & x & z - 6y \end{pmatrix}, \qquad (2.1)$$

where w, x, y, z are complex variables. Two such matrices are of the same type if the numerical coefficients are the same. In this example, the linear combinations of the variables w, x, y, z are taken to be very simple to simplify the numerical computations.

Taking a second matrix in $\mathcal{A}$,

$$P = \begin{pmatrix} 4Y + Z & 4Z - X - 8Y - 4W & 3X & -12Y \\ -Y & W + 2Y & 0 & 3Y \\ 0 & 2Y & W & Y \\ 2Y & 2Z - 4Y - 2W & X & Z - 6Y \end{pmatrix},$$

we note that linear combinations $\lambda Q + \mu P$ also belong to $\mathcal{A}$ because all matrix elements are homogeneous linear functions of w, x, y, z.

For the product QP, matrix multiplication yields

$$QP = \begin{pmatrix} 4y' + z' & 4z' - x' - 8y' - 4w' & 3x' & -12y' \\ -y' & w' + 2y' & 0 & 3y' \\ 0 & 2y' & w' & y' \\ 2y' & 2z' - 4y' - 2w' & x' & z' - 6y' \end{pmatrix},$$

where

$$\begin{aligned} w' &= yX + wW, \\ x' &= xW + zX, \\ y' &= wY + yZ, \\ z' &= xY + zZ. \end{aligned}$$

Thus, the matrix QP is also of type (2.1), which means that $\mathcal{A}$ is an algebra. It is an example of a quantionic algebra. All other isomorphic examples are obtained by similarity transformations

$$Q \mapsto SQS^{-1}, \qquad (2.2)$$

where S is a complex unimodular matrix (meaning $\det S = 1$) which is otherwise arbitrary.

According to the reasoning on pages 11 to 13, the matrix SQS^{-1} contains three conceptually different parts:

(1) The linear part of unification: A matrix

$$A = \begin{pmatrix} w & y \\ x & z \end{pmatrix}$$

has one foot in relativity and the other in quantum mechanics. Specifically, its Hermitian part is related to the linear Minkowski space by the Pauli matrices, while its trace

$$Tr\,(A) = w + z \in \mathbb{C}$$

belongs to the number system of quantum mechanics. Let us denote the algebra of matrices A by $\mathcal{A}_o$. This algebra represents the algebraic part of a structural unification of quantum mechanics and relativity.

(2) The affine part of unification: A unifying number system must be algebraically isomorphic to $\mathcal{A}_o$, but it must also contain additional structure to support differential field equations. The solution — which is beyond guessing — is the algebra $\mathcal{A}$ of complex 4×4 matrices whose matrix elements are linear combinations of the four variables w, x, y, z. A concrete example is the matrix (2.1). The mathematical objects which exist in the algebra $\mathcal{A}$ but not in $\mathcal{A}_o$ are a quantionic derivation operator and the electroweak gauge group $SU\,(2) \times U\,(1)$. The latter is the subgroup of similarity transformations under which the embedding of the algebra $\mathcal{A}_0$ in $\mathcal{A}$ is invariant.

(3) The curvature part of unification: While the search for a quantionic version of the generalization of the affine Minkowski space to a Riemannian space is not undertaken in the present work, let us point out that the algebra of quantions does contain a sufficient number of degrees of freedom to subsume the curvature of spacetime. An example (which may or may not prove to be the source of curvature) is the freedom in the isomorphic embedding of the algebra of 2×2 matrices in the algebra of 4×4 matrices. This freedom is encapsulated in the matrices S of similarity transformations (2.2). Clearly, the simplest 4×4 matrix representation of a 2×2 matrix $A = \begin{pmatrix} w & y \\ x & z \end{pmatrix}$ is the block diagonal form $M = \begin{pmatrix} A & 0 \\ 0 & A \end{pmatrix}$, but an equivalent representation, for example, is the matrix (2.1). Thus, for

$$S = \begin{pmatrix} 0 & -1 & 2 & 0 \\ 1 & 2 & 0 & -1 \\ 0 & 1 & -1 & 0 \\ -1 & 0 & 0 & 2 \end{pmatrix},$$

one easily verifies that

$$S \begin{pmatrix} 4y + z & 4z - x - 8y - 4w & 3x & -12y \\ -y & w + 2y & 0 & 3y \\ 0 & 2y & w & y \\ 2y & 2z - 4y - 2w & x & z - 6y \end{pmatrix} S^{-1} = \begin{pmatrix} w & y & 0 & 0 \\ x & z & 0 & 0 \\ 0 & 0 & w & y \\ 0 & 0 & x & z \end{pmatrix}.$$

Since the matrix S is arbitrary, it may be taken to be a function of spacetime. It is conceivable that this freedom could be a source of differential objects, like affine connections.

2.1   The formal definition of quantions

We begin by introducing some convenient notations.

— The full 16-dimensional algebra of complex 4×4 matrices will be denoted by $\mathcal{M}$.

— The 4-dimensional algebra of quantions, denoted by $\mathbb{D}$ in the Introduction, will be denoted by $\mathcal{L}$ whenever quantions are viewed as complex 4×4 matrices — which is the case throughout the remainder of this book. Thus, $\mathcal{L} \subset \mathcal{M}$.

— "Algebra of quantions" refers to the algebraic properties of the subalgebra $\mathcal{L}$ of $\mathcal{M}$.

— "Theory of quantions" refers to the totality of mathematical objects related to the algebra $\mathcal{L}$.

The present chapter is dedicated to the development of the algebra of quantions. More specifically, to the left version of this algebra.

All concepts introduced in Part I of the present work naturally follow from the following definition:

Definition 1 *The (left) **algebra of quantions**, $\mathcal{L}$, is the algebra of block-diagonal matrices*

$$Q = \begin{pmatrix} A & 0 \\ 0 & A \end{pmatrix}, \tag{2.3}$$

where

$$A = \begin{pmatrix} q_1 & q_3 \\ q_2 & q_4 \end{pmatrix} \tag{2.4}$$

*is an arbitrary complex 2×2 matrix. The matrices (2.3) are referred to as **quantions** (or **L-type quantions** if this is not clear from the context)*

Let us reconsider the physically essential conditions on page 15.

— The associativity of the product of quantions is guaranteed by their matrix representation.

— The complex structure is guaranteed by the complex structure of the complex numbers themselves. Thus, the 'real' and 'imaginary' parts of a quantion are defined as its Hermitian and antihermitian components respectively.

— The polar structure (Chapter 10) and the derivation structure (Chapter 13) are not evident at this point.

— Non-degeneracy has a simple explanation: The quantions have two norm functions, $Q^\dagger Q$ and $\det A$. They both collapse into the single norm function $z^* z$ in the field of complex numbers.

The concepts introduced in this chapter and the next one will be used in this book without repeatedly referring to their definitions.

Definition 2 *A **reduced quantion** is the 2×2 sub-matrix on the diagonal. For an arbitrary quantion $Q \in \mathcal{L}$, we shall denote the reduced quantion by Q_{red}. Thus:*

$$Q \equiv \begin{pmatrix} Q_{red} & 0 \\ 0 & Q_{red} \end{pmatrix}. \tag{2.5}$$

The full algebra of complex 2×2 matrices will be denoted by $\mathcal{L}_{red}$.

All purely algebraic calculations can be done with the reduced quantions if one takes into account that the trace of Q is twice the trace of Q_{red}. We shall see that non-algebraic concepts, like differential equations, require the 4×4 formalism.

Definition 3 *A **Hermitian quantion** is a quantion whose matrix representation is Hermitian: $Q^\dagger = Q$. We shall denote by $\mathcal{L}^h$ the totality of Hermitian quantions.*

The set $\mathcal{L}^h$ of Hermitian quantions is a real linear subspace of $\mathcal{L}$ (because a linear combination of Hermitian matrices is Hermitian if and only if the coefficients are real), but it is not a subalgebra (because the associative product of Hermitian matrices is not a Hermitian matrix).

Definition 4 *A **positive Hermitian quantion** is a Hermitian quantion with a positive-definite trace and a reduced determinant, $\det Q_{red}$, which is not negative:*

$$\left. \begin{array}{rl} Hermiticity: & Q^\dagger = Q, \\ Positive\ trace: & TrQ > 0, \\ Non\text{-}negative\ determinant: & \det Q_{red} \geqslant 0. \end{array} \right\} \qquad (2.6)$$

The totality of positive Hermitian quantions is denoted by $\mathcal{L}^+$.

The following classification of quantions with respect to the value of the determinant plays a key role in physics:

Definition 5 *We shall say that a quantion Q is **regular** if its determinant does not vanish, and that it is **singular** if the determinant does vanish:*

$$\left. \begin{array}{l} Regular\ quantions:\ \det Q \neq 0, \\ Singular\ quantions:\ \det Q = 0. \end{array} \right\} \qquad (2.7)$$

In addition to $\mathcal{L}$ itself, the following subsets of $\mathcal{L}$ will play essential roles in the sequel:

$\mathcal{L}_0$ .... The totality of singular quantions.
$\mathcal{L}_0^h$ .... The totality of singular Hermitian quantions.
$\mathcal{L}_0^+$ ..... The totality of singular positive Hermitian quantions.
$\mathcal{L}^+$ ..... The totality of positive Hermitian quantions.

between covariant and contravariant vectors exists at this point, such a mapping will be derived later as a theorem.

Given an arbitrary quantion

$$
Q = \begin{pmatrix} q_1 & q_3 & 0 & 0 \\ q_2 & q_4 & 0 & 0 \\ 0 & 0 & q_1 & q_3 \\ 0 & 0 & q_2 & q_4 \end{pmatrix} \in \mathcal{L},
\tag{2.19}
$$

one obtains its defining coefficients z^μ from relations (2.16) and (2.17):

$$
\left.\begin{array}{l}
z^0 = \tfrac{1}{4} Tr\,(Q), \\[4pt]
z^1 = \tfrac{1}{4} Tr\,(Q\Lambda_1), \\[4pt]
z^2 = \tfrac{1}{4} Tr\,(Q\Lambda_2), \\[4pt]
z^3 = \tfrac{1}{4} Tr\,(Q\Lambda_3).
\end{array}\right\}
\tag{2.20}
$$

Computation of the right-hand sides yields the relations (2.11).

2.3   The representation space $\mathcal{H}_q$

Since a quantion $P \in \mathcal{L}$ is a complex 4×4 matrix, it may be interpreted as a linear operator acting in a four-dimensional complex linear space. We refer to this auxiliary space as the "quantionic representation space" and denote it by $\mathcal{H}_q$. Thus:

$$
P : \mathcal{H}_q \to \mathcal{H}_q.
$$

Using Dirac's notation, we shall denote the column vectors in $\mathcal{H}_q$ by the ket-like symbol $|*)$,

$$
|q) \in \mathcal{H}_q,
$$

and refer to them as **q-kets**.

By dedicating the unusual symbol $|*)$ to q-kets, we reserve the standard symbol $|*\rangle$ for vectors in the complex Hilbert space and for their generalization to quantionic Hilbert space vectors in any number of dimensions — that is, to vectors whose components are not complex numbers but quantions. (The quantionic Hilbert space is discussed in Chapter 12).

Since quantions and representation vectors are both defined by four independent complex variables, they are linearly isomorphic. It thus follows that the sets $\mathcal{L}$ and $\mathcal{H}_q$ may be related by a one-to-one correspondence

$$\mathcal{L} \ni Q \overset{\omega}{\rightleftarrows} |q) \in \mathcal{H}_q. \tag{2.21}$$

Yet, there is a conceptual difference between a quantion Q and the corresponding q-ket $|q)$, in that the former is an algebraic object (the standard product of matrices is intrinsically defined by the composition of linear transformations) while the latter is only a linear object (there is no *a priori* defined product of vectors, even though an arbitrary product may always be introduced into a linear space by way of a multiplication table).

It remains to be shown that a formal correspondence, which we denote by ω, can be consistently defined within the theory of quantions as a one-to-one mapping between $\mathcal{L}$ and $\mathcal{H}_q$. The consistency condition for the mapping ω is most transparently stated in the form of a commutative diagram:

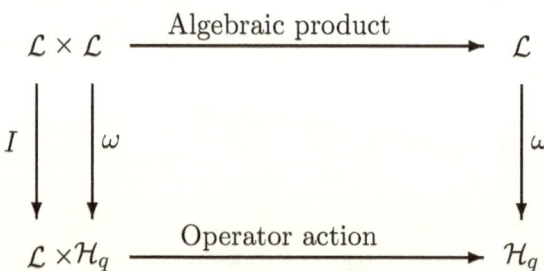

Let us begin with the algebraic product $\mathcal{L} \times \mathcal{L} \to \mathcal{L}$, which is represented by the horizontal arrow at the top of the diagram. Algebraically, it is the product PQ of 4×4 matrices:

$$PQ = \begin{pmatrix} p_1 & p_3 & 0 & 0 \\ p_2 & p_4 & 0 & 0 \\ 0 & 0 & p_1 & p_3 \\ 0 & 0 & p_2 & p_4 \end{pmatrix} \begin{pmatrix} q_1 & q_3 & 0 & 0 \\ q_2 & q_4 & 0 & 0 \\ 0 & 0 & q_1 & q_3 \\ 0 & 0 & q_2 & q_4 \end{pmatrix}$$

$$
= \begin{pmatrix} p_1 q_1 + p_3 q_2 & p_1 q_3 + p_3 q_4 & 0 & 0 \\ p_2 q_1 + p_4 q_2 & p_2 q_3 + p_4 q_4 & 0 & 0 \\ 0 & 0 & p_1 q_1 + p_3 q_2 & p_1 q_3 + p_3 q_4 \\ 0 & 0 & p_2 q_1 + p_4 q_2 & p_2 q_3 + p_4 q_4 \end{pmatrix}.
$$

Turning to the vertical arrows, the identity mapping applies to the quantion P, while the formal definition of the mapping ω remains to be determined. Let us write

$$
\omega : Q \longmapsto |x) = \begin{pmatrix} x_1 \\ x_2 \\ x_3 \\ x_4 \end{pmatrix}.
$$

The operator action $\mathcal{L} \times \mathcal{H}_q \to \mathcal{H}_q$ is represented by the horizontal arrow at the bottom. Algebraically, it is the product $P\,|x)$:

$$
P\,|x) = \begin{pmatrix} p_1 & p_3 & 0 & 0 \\ p_2 & p_4 & 0 & 0 \\ 0 & 0 & p_1 & p_3 \\ 0 & 0 & p_2 & p_4 \end{pmatrix} \begin{pmatrix} x_1 \\ x_2 \\ x_3 \\ x_4 \end{pmatrix} = \begin{pmatrix} p_1 x_1 + p_3 x_2 \\ p_2 x_1 + p_4 x_2 \\ p_1 x_3 + p_3 x_4 \\ p_2 x_3 + p_4 x_4 \end{pmatrix}.
$$

Comparison of the expressions for PQ and $P\,|x)$ yields

$$
|x) = \begin{pmatrix} q_1 \\ q_2 \\ q_3 \\ q_4 \end{pmatrix}.
$$

Thus, the distinguished correspondence ω assumes the simple form

$$
\mathcal{L} \ni Q = \begin{pmatrix} q_1 & q_3 & 0 & 0 \\ q_2 & q_4 & 0 & 0 \\ 0 & 0 & q_1 & q_3 \\ 0 & 0 & q_2 & q_4 \end{pmatrix} \overset{\omega}{\to} \begin{pmatrix} q_1 \\ q_2 \\ q_3 \\ q_4 \end{pmatrix} = |q) \in \mathcal{H}_q. \tag{2.22}
$$

The correspondence (2.22) between quantions and representation vectors will be referred to as **linking**. It admits a compact formalization with the help of a numerical q-ket

$$
|\omega) = \begin{pmatrix} a \\ b \\ c \\ d \end{pmatrix} \in \mathcal{H}_q,
$$

such that

$$\begin{pmatrix} q_1 & q_3 & 0 & 0 \\ q_2 & q_4 & 0 & 0 \\ 0 & 0 & q_1 & q_3 \\ 0 & 0 & q_2 & q_4 \end{pmatrix} \begin{pmatrix} a \\ b \\ c \\ d \end{pmatrix} = \begin{pmatrix} q_1 \\ q_2 \\ q_3 \\ q_4 \end{pmatrix}$$

The solution for $|\omega)$ is unique:

Definition 6 *The **linking vector** is defined as*

$$|\omega) \stackrel{def}{=} \begin{pmatrix} 1 \\ 0 \\ 0 \\ 1 \end{pmatrix}. \tag{2.23}$$

The vector $|q) \in \mathcal{H}_q$ linked to an arbitrary quantion $Q \in \mathcal{L}$ is thus given by the relation

$$|q) = Q\,|\omega). \tag{2.24}$$

It is the sum of the first and last columns of Q.

Taking the Hermitian conjugate of relation (2.24), one obtains the q-bra

$$(q| = (\omega|\,Q^\dagger. \tag{2.25}$$

Explicitly:

$$\begin{aligned} (q| &= \begin{pmatrix} 1 & 0 & 0 & 1 \end{pmatrix} \begin{pmatrix} q_1^* & q_2^* & 0 & 0 \\ q_3^* & q_4^* & 0 & 0 \\ 0 & 0 & q_1^* & q_3^* \\ 0 & 0 & q_3^* & q_4^* \end{pmatrix} \\ &= \begin{pmatrix} q_1^* & q_2^* & q_3^* & q_4^* \end{pmatrix} \\ &= [|q)]^\dagger. \end{aligned} \tag{2.26}$$

Since both a q-bra and a q-ket are uniquely linked to every quantion, the scalar product

$$(p|q) \stackrel{def}{=} p_1^* q_1 + p_2^* q_2 + p_3^* q_3 + p_4^* q_4 \in \mathbb{C} \tag{2.27}$$

exists unconditionally and is structurally defined. The representation $\mathcal{H}_q$ may thus be viewed as a four-dimensional Hilbert space.

In term of the linking vector ω, the trace of a quantion assumes the compact form

$$TrQ = 2\left(\omega|Q|\omega\right). \tag{2.28}$$

Clearly, this expression is valid only for quantions — not for arbitrary matrices $M \in \mathcal{M}$.

The relations (2.20) may now be rewritten in the elegant form

$$\left.\begin{aligned}
z^0 &= \tfrac{1}{2}\left(\omega|Q|\omega\right), \\
z^1 &= \tfrac{1}{2}\left(\omega|Q\Lambda_1|\omega\right), \\
z^2 &= \tfrac{1}{2}\left(\omega|Q\Lambda_2|\omega\right), \\
z^3 &= \tfrac{1}{2}\left(\omega|Q\Lambda_3|\omega\right).
\end{aligned}\right\} \tag{2.29}$$

The linking vector $|\omega)$ maps the algebra $\mathcal{L}$ of quantions onto the representation space $\mathcal{H}_q$ without loss of information. Thus, a mapping inverse to (2.22) also exists. Let us denote it by Ω :

$$\begin{pmatrix} q_1 \\ q_2 \\ q_3 \\ q_4 \end{pmatrix} \xrightarrow{\Omega} \begin{pmatrix} q_1 & q_3 & 0 & 0 \\ q_2 & q_4 & 0 & 0 \\ 0 & 0 & q_1 & q_3 \\ 0 & 0 & q_2 & q_4 \end{pmatrix}.$$

To define the mapping Ω formally, we introduce the four numerical matrices that form a linear basis in the algebra $\mathcal{L}$ when the coefficients are q_1 to q_4. These matrices can be read off the matrix Q in relation (2.19):

$$M_1 \stackrel{def}{=} \begin{pmatrix} 1 & 0 & 0 & 0 \\ 0 & 0 & 0 & 0 \\ 0 & 0 & 1 & 0 \\ 0 & 0 & 0 & 0 \end{pmatrix} = \frac{1}{2}\left(\Lambda_0 + \Lambda_3\right), \tag{2.30}$$

$$M_2 \stackrel{def}{=} \begin{pmatrix} 0 & 0 & 0 & 0 \\ 1 & 0 & 0 & 0 \\ 0 & 0 & 0 & 0 \\ 0 & 0 & 1 & 0 \end{pmatrix} = \frac{1}{2}\left(\Lambda_1 - i\Lambda_2\right), \tag{2.31}$$

$$M_3 \stackrel{def}{=} \begin{pmatrix} 0 & 1 & 0 & 0 \\ 0 & 0 & 0 & 0 \\ 0 & 0 & 0 & 1 \\ 0 & 0 & 0 & 0 \end{pmatrix} = \frac{1}{2}\left(\Lambda_1 + i\Lambda_2\right), \tag{2.32}$$

$$M_4 \stackrel{def}{=} \begin{pmatrix} 0 & 0 & 0 & 0 \\ 0 & 1 & 0 & 0 \\ 0 & 0 & 0 & 0 \\ 0 & 0 & 0 & 1 \end{pmatrix} = \frac{1}{2}\left(\Lambda_0 - \Lambda_3\right). \tag{2.33}$$

Given a q-ket $|q)$, the corresponding quantion Q is thus

$$Q = q_1 M_1 + q_2 M_2 + q_3 M_3 + q_4 M_4. \tag{2.34}$$

This linear combination may be compactly written in the form of a scalar product involving the q-ket $|q)$:

$$Q = \begin{pmatrix} M_1 & M_2 & M_3 & M_4 \end{pmatrix} \begin{pmatrix} q_1 \\ q_2 \\ q_3 \\ q_4 \end{pmatrix}.$$

Thus, the **inverse linking vector** is formally a q-bra whose components are not numbers but matrices:

$$(\Omega| \stackrel{def}{=} \begin{pmatrix} M_1 & M_2 & M_3 & M_4 \end{pmatrix}. \tag{2.35}$$

The mapping

$$\mathcal{H}_q \stackrel{\Omega}{\to} \mathcal{L}$$

now assumes the compact form

$$Q = (\Omega|q). \tag{2.36}$$

The mappings Ω and ω performed in succession yield a projection operator from $\mathcal{M}$ onto $\mathcal{L}$. Thus, for an arbitrary matrix $M \in \mathcal{M}$, the matrix

$$\hat{M} = (\Omega|M|\omega) \in \mathcal{L} \tag{2.37}$$

is a quantion.

2.4 The concept of linking

The concept referred to above a "linking" is not restricted to the algebra of quantions. It relates any associative algebra to its (left) regular representation. Yet, to the best of the author's knowledge, linking is not a standard mathematical concept under any name (which would probably be other than "linking" if it were to be found in the literature). This tells us that it is a concept which has not been sufficiently useful in mathematics to deserve a name. Its usefulness in the theory of quantions is not evident at this point either. It will become evident only in quantionic differential equations, as pointed out at the end of Chapter 13.

The objectives of the present section are:

(a) to put the idea of linking in perspective by defining it over any associative algebra, and

(b) to introduce this idea into the field of complex numbers in preparation for Chapter 13.

Linking in arbitrary associative algebras

We shall begin with a brief overview of the standard concept of a regular representation.

Let $\mathcal{H}$ be a real or complex linear space of some dimension n. Further, let an associative product $\circ$ be defined in $\mathcal{H}$:

$$\circ : \mathcal{H} \times \mathcal{H} \to \mathcal{H}.$$

In other words $\{\mathcal{H}, \circ\}$ is an associative algebra defined by some multiplication table. Thus, for every pair of vectors

$$|p) , |q) \in \mathcal{H}$$

there exists a unique vector $|u) \in \mathcal{H}$ defined as

$$|u) = |p) \circ |q) .$$

Since the product $\circ$ is associative, there exists a uniquely defined $n \times n$ matrix P such that

$$P |q) = |p) \circ |q)$$

the arithmetic of complex numbers follows as a theorem from the algebra of matrices:

$$WZ = \begin{pmatrix} u & -v \\ v & u \end{pmatrix} \begin{pmatrix} x & -y \\ y & x \end{pmatrix} = \begin{pmatrix} ux - vy & -(uy + vx) \\ uy + vx & ux - vy \end{pmatrix}.$$

To obtain the linking vector for complex numbers, we write

$$|\omega) = \begin{pmatrix} \xi \\ \eta \end{pmatrix}$$

and impose the defining condition $z = Z\,|\omega)$, that is:

$$\begin{pmatrix} x \\ y \end{pmatrix} = \begin{pmatrix} x & -y \\ y & x \end{pmatrix} \begin{pmatrix} \xi \\ \eta \end{pmatrix} = \begin{pmatrix} x\xi - y\eta \\ x\eta + y\xi \end{pmatrix}.$$

The solution is

$$|\omega) = \begin{pmatrix} 1 \\ 0 \end{pmatrix}.$$

Other concepts related to complex numbers follow readily from their matrix representation:

Complex conjugation corresponds to transposition:

$$z^* = x - iy \rightleftarrows Z^{\mathsf{T}} = \begin{pmatrix} x & y \\ -y & x \end{pmatrix}.$$

The real and imaginary units correspond, respectively, to the unit matrix and the matrix

$$J = \begin{pmatrix} 0 & -1 \\ 1 & 0 \end{pmatrix}.$$

Thus:

$$z = x + iy \rightleftarrows Z = xI + yJ.$$

Finally, the norm

$$\|z\|^2 \stackrel{def}{=} z^*z = x^2 + y^2$$

of a complex number is

$$Z^{\mathsf{T}} Z = \begin{pmatrix} x & y \\ -y & x \end{pmatrix} \begin{pmatrix} x & -y \\ y & x \end{pmatrix} = (x^2 + y^2)\, I.$$

Chapter 3

The Right Algebra of Quantions

To the left algebra $\mathcal{L} \subset \mathcal{M}$ of quantions is uniquely associated its commutant, which is a second subalgebra, $\mathcal{R} \subset \mathcal{M}$. We refer to it as the "right algebra of quantions". The mutual relationshps of the algebras $\mathcal{L}$ and $\mathcal{R}$ and of the representation space $\mathcal{H}_q$ generate several new concepts that play key roles in the structural unification developed in Part II. These concepts are bought to light in the present chapter.

3.1 The commutants

The **commutant** of the subalgebra $\mathcal{L}$ of $\mathcal{M}$ is defined as the subset $\mathcal{R}$ of all matrices in $\mathcal{M}$ that commute with every matrix in $\mathcal{L}$. Symbolically,

$$[\mathcal{L}, \mathcal{R}] = \{0\} .$$

To verify that $\mathcal{R}$ is also an algebra, let Q be an arbitrary element of $\mathcal{L}$, and let R and S be arbitrary elements of $\mathcal{R}$. Then, by definition, $[Q, R] = [Q, S] = 0$. It follows that $[Q, \lambda R + S] = 0$ for every complex number λ, which implies that $\mathcal{R}$ is a complex linear space. On the other hand, the general commutator identity

$$[Q, RS] \equiv [Q, R] S + R [Q, S]$$

41

implies $[Q, RS] = 0$, that is, $RS \in \mathcal{R}$. Hence, $\mathcal{R}$ is also an algebra (an associative algebra, for being an algebra of matrices.)

The most general matrices in $\mathcal{R}$ are of the form given by the following theorem:

Theorem 7 *The matrices of the algebra $\mathcal{R}$ are of the type*

$$R = \begin{pmatrix} fI & hI \\ gI & kI \end{pmatrix} = \begin{pmatrix} f & 0 & h & 0 \\ 0 & f & 0 & h \\ g & 0 & k & 0 \\ 0 & g & 0 & k \end{pmatrix}, \qquad (3.1)$$

where f, g, h, k are arbitrary complex numbers.

Proof. Let L be an arbitrary quantion,

$$L = \begin{pmatrix} A & 0 \\ 0 & A \end{pmatrix} \in \mathcal{L},$$

and $M \in \mathcal{M}$ an arbitrary complex 4×4 matrix. We may write M in block form,

$$M = \begin{pmatrix} F & H \\ G & K \end{pmatrix},$$

where $F, G, H,$ and K are arbitrary complex 2×2 matrices. The matrix M is in the commutant $\mathcal{R}$ of $\mathcal{L}$ if it satisfies the commutativity condition

$$[M, L] = \begin{pmatrix} [F, A] & [H, A] \\ [G, A] & [K, A] \end{pmatrix} = 0$$

This relation implies that the four matrices F, G, H, K commute with the 2×2 matrix A. The latter being arbitrary, it follows that they are all multiples of the unit matrix, that is, $F = fI$, $G = gI$, $H = hI$, $K = kI$, for arbitrary complex coefficients f, g, h, k. This conclusion yields the most general expression (3.1) for the matrices in $\mathcal{R}$. ∎

The **center** of any algebra $\mathcal{A}$ is defined as the totality of matrices in $\mathcal{A}$ that commute with every matrix in $\mathcal{A}$. In the algebras of quantions, this definition yields the following theorem:

Theorem 8 *The algebras $\mathcal{L}$ and $\mathcal{R}$ have a common center. It is the field $\mathbb{C}$ of complex numbers:*

$$\mathcal{L} \cap \mathcal{R} = \mathbb{C}I. \tag{3.2}$$

Proof. The intersection $\mathcal{L} \cap \mathcal{R}$ is defined by the condition

$$\mathcal{L} \ni \begin{pmatrix} A & 0 \\ 0 & A \end{pmatrix} = \begin{pmatrix} fI & hI \\ gI & kI \end{pmatrix} \in \mathcal{R},$$

which implies $g = h = 0$ and $A = fI = kI$. It follows that the common substructure is essentially the field $\mathbb{C}$ of complex numbers written in the form $\mathbb{C}I$. This field is also the center of each of the two algebras because only the complex numbers commute with all 2×2 matrices Q_{red}. ■

3.2 Algebraic duality

The subalgebras $\mathcal{L}$ and $\mathcal{R}$ are mutually related by a structurally distinguished one-to-one correspondence. We refer to it as "algebraic duality". Unless obvious from the context, it is essential to specify this duality as "algebraic" to avoid confusion with the "metric" duality that will be introduced in Section 6.1.

The underlying idea is the observation that the linking

$$\mathcal{L} \ni Q \rightleftharpoons |q) \in \mathcal{H}_q$$

is established by the condition

$$PQ \rightleftharpoons P|q), \tag{3.3}$$

in which the quantion $P \in \mathcal{L}$, interpreted as a linear operator, acts on q-kets from the left (which justifies the terminology "left quantion"). But this is an arbitrary decision. The opposite decision,

$$PQ \rightleftharpoons (p^*|Q, \tag{3.4}$$

would be equally justified, the vectors $(p^*|$ and $|p)$ being related by transposition:

$$\begin{pmatrix} p_1 & p_2 & p_3 & p_4 \end{pmatrix} = (p^*| \underset{\rightleftharpoons}{\overset{\top}{}} |p) = \begin{pmatrix} p_1 \\ p_2 \\ p_3 \\ p_4 \end{pmatrix}. \tag{3.5}$$

Since the left-hand sides of (3.3) and (3.4) coincide, the right-hand sides ought to be identical as well, but this not possible in the given forms for two reasons:

— While $P\,|q)$ is a q-ket, $(p^*|\,Q$ is a q-bra. This is easily adjusted by taking the transpose of one of these vectors, let's say $[P\,|q)]^\top$, instead of $P\,|q)$.

— Since the linking $Q \rightleftarrows |q)$ has already been established by the relation $Q = (\Omega|q)$, it cannot be redefined it by the relation (3.4). We can, however, define a new matrix $\tilde{Q}$ that acts on q-bras from the right: $(p^*|\,\tilde{Q}$.

Combining these observations, we introduce the following definition of algebraic duality:

Definition 9 *Two matrices, $Q \in \mathcal{L}$ and $\tilde{Q} \in \mathcal{R}$, are said to be **algebraically dual** to each other if the relation*

$$(p^*|\,\tilde{Q} \equiv [P\,|q)]^\top \tag{3.6}$$

holds, the identity being with respect to all $P \in \mathcal{L}$.

The following theorem explicitly relates the matrix expressions of algebraically dual quantions.

Theorem 10 *Given a left quantion $Q \in \mathcal{L}$,*

$$Q = \begin{pmatrix} q_1 & q_3 & 0 & 0 \\ q_2 & q_4 & 0 & 0 \\ 0 & 0 & q_1 & q_3 \\ 0 & 0 & q_2 & q_4 \end{pmatrix},$$

its algebraic dual $\tilde{Q} \in \mathcal{R}$ is the right quantion

$$\tilde{Q} = \begin{pmatrix} q_1 & 0 & q_3 & 0 \\ 0 & q_1 & 0 & q_3 \\ q_2 & 0 & q_4 & 0 \\ 0 & q_2 & 0 & q_4 \end{pmatrix}, \tag{3.7}$$

and vice-versa.

Proof. Let us write the unknown matrix $\tilde{Q} \in \mathcal{M}$ in the form

$$\tilde{Q} = \begin{pmatrix} a & 0 & b & 0 \\ 0 & c & 0 & d \\ f & 0 & g & 0 \\ 0 & h & 0 & k \end{pmatrix} + \begin{pmatrix} 0 & \alpha & 0 & \beta \\ \gamma & 0 & \delta & 0 \\ 0 & \lambda & 0 & \mu \\ \rho & 0 & \sigma & 0 \end{pmatrix}. \tag{3.8}$$

Hence, the left-hand side of relation (3.6) reads

$$
\begin{aligned}
(p^* | \tilde{Q} &= \begin{pmatrix} p_1 & p_2 & p_3 & p_4 \end{pmatrix} \begin{pmatrix} a & 0 & b & 0 \\ 0 & c & 0 & d \\ f & 0 & g & 0 \\ 0 & h & 0 & k \end{pmatrix} \\
&+ \begin{pmatrix} p_1 & p_2 & p_3 & p_4 \end{pmatrix} \begin{pmatrix} 0 & \alpha & 0 & \beta \\ \gamma & 0 & \delta & 0 \\ 0 & \lambda & 0 & \mu \\ \rho & 0 & \sigma & 0 \end{pmatrix} \\
&= \begin{pmatrix} ap_1 + fp_3 & cp_2 + hp_4 & bp_1 + gp_3 & dp_2 + kp_4 \end{pmatrix} \\
&+ \begin{pmatrix} \gamma p_2 + \rho p_4 & \alpha p_1 + \lambda p_3 & \sigma p_4 + \delta p_2 & \beta p_1 + \mu p_3 \end{pmatrix}.
\end{aligned}
$$

The right-hand side of relation (3.6) is

$$
\begin{aligned}
[P \,|\, q)]^\top &= \left\{ \begin{pmatrix} p_1 & p_3 & 0 & 0 \\ p_2 & p_4 & 0 & 0 \\ 0 & 0 & p_1 & p_3 \\ 0 & 0 & p_2 & p_4 \end{pmatrix} \begin{pmatrix} q_1 \\ q_2 \\ q_3 \\ q_4 \end{pmatrix} \right\}^\top = \begin{pmatrix} p_1 q_1 + p_3 q_2 \\ p_2 q_1 + p_4 q_2 \\ p_1 q_3 + p_3 q_4 \\ p_2 q_3 + p_4 q_4 \end{pmatrix}^\top \\
&= \begin{pmatrix} p_1 q_1 + p_3 q_2 & p_2 q_1 + p_4 q_2 & p_1 q_3 + p_3 q_4 & p_2 q_3 + p_4 q_4 \end{pmatrix}.
\end{aligned}
$$

The identity condition (3.6) eliminates the variables p_i and yields

$$
\begin{aligned}
\alpha &= \beta = \gamma = \delta = 0, \\
a &= c = q_1, \\
f &= h = q_2, \\
b &= d = q_3, \\
g &= k = q_4.
\end{aligned}
$$

Substitution of these results into the matrices (3.8) yields the matrix type (3.7). ∎

It is now evident that the terminology "left" and "right" for the two algebras of quantions is justified by the identity (3.6). In this identity, the L-type quantion $Q \in \mathcal{L}$ acts on $\mathcal{H}_q$ from the left, the R-type quantion $\tilde{Q} \in \mathcal{R}$ acts from the right.

The duality Theorem 9 extends the linking relation (2.24) to $\mathcal{R}$:

Theorem 11 *L-type quantions are linked to q-kets, R-type quantions to q-bras:*

$$Q \left| \omega \right) = \left| q \right) \overset{\top}{\rightleftarrows} \left(q^* \right| = \left(\omega \right| \tilde{Q}. \tag{3.9}$$

Proof. The right-hand side

$$\left(q^* \right| = \left(\omega \right| \tilde{Q} = \begin{pmatrix} 1 & 0 & 0 & 1 \end{pmatrix} \begin{pmatrix} q_1 & 0 & q_3 & 0 \\ 0 & q_1 & 0 & q_3 \\ q_2 & 0 & q_4 & 0 \\ 0 & q_2 & 0 & q_4 \end{pmatrix} = \begin{pmatrix} q_1 & q_2 & q_3 & q_4 \end{pmatrix}$$

is indeed the transpose of $\left| q \right)$. ∎

An observation on terminology: Depending on the properties one wishes to emphasize, the algebras $\mathcal{L}$ and $\mathcal{R}$ may be referred to as commutants, algebraic duals, left and right algebras of quantions, or left and right regular representations of the algebra of complex 2×2 matrices written in terms of quadruples:

$$\{a, b, c, d\} \{u, v, w, z\} = \{au + cv, bu + dv, aw + cz, bw + dz\} .$$

The symbols $\mathcal{R}^h$, $\mathcal{R}^+$, $\mathcal{R}_0$, and $\mathcal{R}_0^+$ will be used as the R-type counterparts for the symbols $\mathcal{L}^h$, $\mathcal{L}^+$, $\mathcal{L}_0$, and $\mathcal{L}_0^+$ introduced in Chapter 2 for the sets of Hermitian, positive Hermitian, singular, and positive singular L-type quantions respectively.

Definition 12 *The reduced form of the R-type quantion*

$$\tilde{Q} = \begin{pmatrix} q_1 I & q_3 I \\ q_2 I & q_4 I \end{pmatrix}$$

is the 2×2 complex matrix

$$\tilde{Q}_{red} = \begin{pmatrix} q_1 & q_3 \\ q_2 & q_4 \end{pmatrix}. \tag{3.10}$$

The algebra of matrices $\tilde{Q}_{red}$ will be denoted by $\mathcal{R}_{red}$. Clearly:

$$\mathcal{R}_{red} = \mathcal{L}_{red}. \tag{3.11}$$

The distinction between left and right quantions is thus lost in the transition to the purely algebraic reduced version of their algebra.

3.3 The similarity transformation W

Theorem 13 *The mutually dual matrices $Q \in \mathcal{L}$ and $\tilde{Q} \in \mathcal{R}$, which are related by Theorem 10, are also related by a similarity transformation*

$$\tilde{Q} = WQW^{-1}, \tag{3.12}$$

where

$$W \stackrel{def}{=} \begin{pmatrix} 1 & 0 & 0 & 0 \\ 0 & 0 & 1 & 0 \\ 0 & 1 & 0 & 0 \\ 0 & 0 & 0 & 1 \end{pmatrix} = W^{-1}. \tag{3.13}$$

Proof. If a similarity transformation $\tilde{Q} = SQS^{-1}$ exists, it may also be written as

$$\tilde{Q}S = SQ. \tag{3.14}$$

Taking for the unknown S the most general 4×4 matrix

$$S = \begin{pmatrix} a & b & c & d \\ f & g & h & k \\ m & n & p & r \\ u & v & w & z \end{pmatrix}.$$

we compute both sides of condition (3.14) independently. For the left-hand side, one obtains

$$\tilde{Q}S = \begin{pmatrix} q_1 & 0 & q_3 & 0 \\ 0 & q_1 & 0 & q_3 \\ q_2 & 0 & q_4 & 0 \\ 0 & q_2 & 0 & q_4 \end{pmatrix} \begin{pmatrix} a & b & c & d \\ f & g & h & k \\ m & n & p & r \\ u & v & w & z \end{pmatrix}$$

$$= \begin{pmatrix} aq_1 + mq_3 & bq_1 + nq_3 & cq_1 + pq_3 & dq_1 + rq_3 \\ fq_1 + uq_3 & gq_1 + vq_3 & hq_1 + wq_3 & kq_1 + zq_3 \\ aq_2 + mq_4 & bq_2 + nq_4 & cq_2 + pq_4 & dq_2 + rq_4 \\ fq_2 + uq_4 & gq_2 + vq_4 & hq_2 + wq_4 & kq_2 + zq_4 \end{pmatrix}.$$

The right-hand side is

$$SQ = \begin{pmatrix} a & b & c & d \\ f & g & h & k \\ m & n & p & r \\ u & v & w & z \end{pmatrix} \begin{pmatrix} q_1 & q_3 & 0 & 0 \\ q_2 & q_4 & 0 & 0 \\ 0 & 0 & q_1 & q_3 \\ 0 & 0 & q_2 & q_4 \end{pmatrix}$$

$$= \begin{pmatrix} aq_1 + bq_2 & aq_3 + bq_4 & cq_1 + dq_2 & cq_3 + dq_4 \\ fq_1 + gq_2 & fq_3 + gq_4 & hq_1 + kq_2 & hq_3 + kq_4 \\ mq_1 + nq_2 & mq_3 + nq_4 & pq_1 + rq_2 & pq_3 + rq_4 \\ uq_1 + vq_2 & uq_3 + vq_4 & wq_1 + zq_2 & wq_3 + zq_4 \end{pmatrix}.$$

The similarity condition (3.14) is to be understood as an identity in the independent variables q_1 to q_4 and as a system of equations for the sixteen unknowns a to z. Hence, elimination of the variables q_1 to q_4 yields the following system of equations:

$$\begin{aligned} m &= b = p = d = u = g = w = k = 0, \\ n &= a, \\ v &= f, \\ c &= r, \\ h &= z. \end{aligned}$$

The most general solution for S is thus

$$S = \begin{pmatrix} a & 0 & c & 0 \\ v & 0 & h & 0 \\ 0 & a & 0 & c \\ 0 & v & 0 & h \end{pmatrix} \tag{3.15}$$

under the condition

$$\det S = \det \begin{pmatrix} a & 0 & c & 0 \\ v & 0 & h & 0 \\ 0 & a & 0 & c \\ 0 & v & 0 & h \end{pmatrix} = -(ah - cv)^2 \neq 0. \tag{3.16}$$

Since the matrix S is defined up to an arbitrary complex factor, the condition (3.16) may be normalized to

$$(ah - cv)^2 = 1, \tag{3.17}$$

in which case the inverse of S is

$$S^{-1} = \begin{pmatrix} h & -c & 0 & 0 \\ 0 & 0 & h & -c \\ -v & a & 0 & 0 \\ 0 & 0 & -v & a \end{pmatrix}. \tag{3.18}$$

We now observe that if a matrix S satisfies the similarity condition $\tilde{Q} = SQS^{-1}$, so does the matrix SR for an arbitrary non-singular R-type quantion $R \in \mathcal{R}$. The reason is that R and Q commute. We thus have the sequence of equation

$$\begin{aligned} (SR)\, Q\, (SR)^{-1} &= SRQR^{-1}S^{-1} \\ &= SQRR^{-1}S^{-1} = SQS^{-1} = \tilde{Q}. \end{aligned}$$

For convenience, we may select a similarity matrix distinguished by a property that suggests itself, namely $S^{-1} = S$. Equating the matrices (3.15) and (3.18) yields

$$\begin{aligned} a &= h = 1, \\ v &= c = 0. \end{aligned}$$

It follows that the solution for such a matrix S is the matrix W given by (3.13).

It remains to verify that we may work with the matrix W without loss of generality, that is, the matrix W is not less general than the matrix (3.15). To this end, let R be an arbitrary R-type quantion

$$R = \begin{pmatrix} a & 0 & c & 0 \\ 0 & a & 0 & c \\ v & 0 & h & 0 \\ 0 & v & 0 & h \end{pmatrix} \in \mathcal{R}.$$

The matrix WR is then

$$WR = \begin{pmatrix} 1 & 0 & 0 & 0 \\ 0 & 0 & 1 & 0 \\ 0 & 1 & 0 & 0 \\ 0 & 0 & 0 & 1 \end{pmatrix} \begin{pmatrix} a & 0 & c & 0 \\ 0 & a & 0 & c \\ v & 0 & h & 0 \\ 0 & v & 0 & h \end{pmatrix} = \begin{pmatrix} a & 0 & c & 0 \\ v & 0 & h & 0 \\ 0 & a & 0 & c \\ 0 & v & 0 & h \end{pmatrix},$$

which is indeed the general solution for S given by relation (3.15). This proves the theorem. ■

We shall refer to W as the **duality matrix.**

Theorem 14 *To the linking relation (2.24), which maps the left algebra $\mathcal{L}$ onto the kets in the Hilbert space $\mathcal{H}_q$, corresponds the linking relation*

$$(q^*| = (\omega|\,\tilde{Q}, \tag{3.19}$$

which maps the right algebra $\mathcal{R}$ onto the bras in $\mathcal{H}_q$.

Proof. Substitution of the expressions for $(\omega|$ and $\tilde{Q}$ yields

$$(\omega|\,\tilde{Q} = \begin{pmatrix} 1 & 0 & 0 & 1 \end{pmatrix} \begin{pmatrix} q_1 & 0 & q_3 & 0 \\ 0 & q_1 & 0 & q_3 \\ q_2 & 0 & q_4 & 0 \\ 0 & q_2 & 0 & q_4 \end{pmatrix}$$

$$= \begin{pmatrix} q_1 & q_2 & q_3 & q_4 \end{pmatrix} = (q^*|\,,$$

which proves the assertion. ■

3.4   An equivalence relation in $\mathcal{M}$

In the linking relation

$$|q) = Q\,|\omega)\,, \tag{3.20}$$

the q-ket $|q) \in \mathcal{H}_q$ is uniquely defined not only for the L-type quantions $Q \in \mathcal{L}$, but for all matrices $Q \in \mathcal{M}$. But since the linear spaces $\mathcal{H}_q$ and $\mathcal{M}$ have four and sixteen complex dimensions respectively, it follows that the mapping

$$\omega : \mathcal{M} \to \mathcal{H}_q \tag{3.21}$$

assigns the same q-ket to all matrices $M \in \mathcal{M}$ for which the sum of the first and last columns are identical. Thus, given a q-ket $|q)$, the most general solution of the equation $M\,|\omega) = |q)$ is

$$M = \frac{1}{2}\begin{pmatrix} q_1 + z_1 & z_5 & z_9 & q_1 - z_1 \\ q_2 + z_2 & z_6 & z_{10} & q_2 - z_2 \\ q_3 + z_3 & z_7 & z_{11} & q_3 - z_3 \\ q_4 + z_4 & z_8 & z_{12} & q_4 - z_4 \end{pmatrix} \in \mathcal{M}. \tag{3.22}$$

In this matrix, the twelve complex numbers z_1 to z_{12} are arbitrary. For every $|q)$, the totality of these matrices is a 12-dimensional subspace of $\mathcal{M}$ which defines an equivalence relation:

Definition 15 *We shall say that two matrices $A, B \in \mathcal{M}$ are equivalent, denoted by $A \sim B$, if they map the vector $|\omega)$ onto the same q-ket in $\mathcal{H}_q$:*

$$A\,|\omega) = B\,|\omega). \tag{3.23}$$

*We shall refer to the equivalence "$\sim$" as the **omega-equivalence**.*

Thus, the linear subspace of matrices (3.22) is an equivalence class with respect to the omega-equivalence relation.

Let $\mathcal{Z}$ denote the kernel of the mapping $\omega : \mathcal{M} \to \mathcal{H}_q$, that is, the 12-dimensional linear subspace $\mathcal{Z} \subset \mathcal{M}$ of matrices whose image in $\mathcal{L}$ is zero,

$$\mathcal{Z}\,|\omega) = 0. \tag{3.24}$$

The elements of $\mathcal{Z}$ are the matrices M_0 of the type

$$M_0 = \begin{pmatrix} z_1 & z_5 & z_9 & -z_1 \\ z_2 & z_6 & z_{10} & -z_2 \\ z_3 & z_7 & z_{11} & -z_3 \\ z_4 & z_8 & z_{12} & -z_4 \end{pmatrix} \in \mathcal{Z}. \tag{3.25}$$

Then, a more elegant way of defining the omega-equivalence condition (3.23) is to say that two matrices $A, B \in \mathcal{M}$ are equivalent if and only if their difference is in the kernel $\mathcal{Z}$,

$$(A - B) \in \mathcal{Z}. \tag{3.26}$$

It is thus evident that every quantion $Q \in \mathcal{L}$ belongs to one and only one equivalence class, $\mathcal{S}\,(Q) \subset \mathcal{M}$, defined as

$$\mathcal{S}\,(Q) = Q + \mathcal{Z}, \tag{3.27}$$

which means that Q is to be added to every element of $\mathcal{Z}$.

The definition 15 of omega-equivalence is with respect to the left algebra $\mathcal{L}$ of quantions, but a similar equivalence could de defined, should it prove useful, with respect to the right algebra $\mathcal{R}$.

3.5   The pi matrices

In Section 2.2, we introduced a system of basis matrices in the algebra $\mathcal{L}$ — the lambda matrices . In this section, we introduce a similar basis in the algebra $\mathcal{R}$. We refer to as the pi matrices.

The lambda matrices were selected as the simplest 4×4 representation of the Pauli sigma matrices — both for computational convenience and as a link to the standard formalism of quantum field theory. But once a basis has been selected in $\mathcal{L}$, a basis in $\mathcal{R}$ is uniquely defined by algebraic duality.

Let Q be an arbitrary L-type quantion expanded in the basis of lambda matrices, $Q = z^\mu \Lambda_\mu$. Its algebraic dual, $\tilde{Q}$, is defined by the relation (3.12). Thus,

$$\tilde{Q} = WQW = W\left(z^\mu \Lambda_\mu\right)W.$$

Since the coefficients z^μ are complex numbers, they commute with the similarity matrix W. Hence, denoting by Π_μ the algebraic duals of the matrices Λ_μ,

$$\Pi_\mu \overset{def}{=} W\Lambda_\mu W, \tag{3.28}$$

the expansion of the algebraic dual $\tilde{Q} \in \mathcal{R}$ of $Q \in \mathcal{L}$ reads

$$\tilde{Q} = z^\mu \Pi_\mu. \tag{3.29}$$

It follows that the matrices Π_μ form a basis in the algebra $\mathcal{R}$.

Since the matrix W has no convenient block form (because none of the four 2×2 blocks vanishes), the right-hand side of (3.28) must be computed in the 4×4 formalism. The products have nevertheless very simple expressions in bock form, the blocks being proportional to the 2×2 unit matrix I :

$$\Pi_0 = \Lambda_0 = I = \begin{pmatrix} 1 & 0 & 0 & 0 \\ 0 & 1 & 0 & 0 \\ 0 & 0 & 1 & 0 \\ 0 & 0 & 0 & 1 \end{pmatrix} = \begin{pmatrix} I & 0 \\ 0 & I \end{pmatrix}, \tag{3.30}$$

$$\Pi_1 = W\Lambda_1 W = \begin{pmatrix} 0 & 0 & 1 & 0 \\ 0 & 0 & 0 & 1 \\ 1 & 0 & 0 & 0 \\ 0 & 1 & 0 & 0 \end{pmatrix} = \begin{pmatrix} 0 & I \\ I & 0 \end{pmatrix}, \tag{3.31}$$

$$\Pi_2 = W\Lambda_2 W = \begin{pmatrix} 0 & 0 & -i & 0 \\ 0 & 0 & 0 & -i \\ i & 0 & 0 & 0 \\ 0 & i & 0 & 0 \end{pmatrix} = \begin{pmatrix} 0 & -iI \\ iI & 0 \end{pmatrix}, \tag{3.32}$$

$$\Pi_3 = W\Lambda_3 W = \begin{pmatrix} 1 & 0 & 0 & 0 \\ 0 & 1 & 0 & 0 \\ 0 & 0 & -1 & 0 \\ 0 & 0 & 0 & -1 \end{pmatrix} = \begin{pmatrix} I & 0 \\ 0 & -I \end{pmatrix}. \tag{3.33}$$

Theorem 16 *The multiplication table of pi matrices is structurally identical to the multiplication table of lambda matrices.*

Proof. This is trivially true for all products in which one of the factors is $\Pi_0 = I$. Let us consider the products of the matrices Π_i.
The case of $i \neq j$:

$$\Pi_i \Pi_j = (W\Lambda_i W)(W\Lambda_j W) = W\Lambda_i W^2 \Lambda_j W.$$

Since $W^2 = I$, we have

$$\Pi_i \Pi_j = W\Lambda_i \Lambda_j W = \varepsilon_{ijk} i W\Lambda_k W.$$

Hence, if the indices i, j, k are taken in the positive cyclic order, this relation simplifies to $\Pi_i \Pi_j = i\Pi_k$.
The case of $i = j$:

$$\Pi_i^2 = W\Lambda_i^2 W = W^2 = I.$$

Finally, we see from the expressions (3.30) to (3.33) that the only matrix pi whose trace does not vanish is Π_0. Hence the properties of the pi matrices are:

$$\left.\begin{array}{l} \Pi_\mu^2 = I, \\ \Pi_i \Pi_j = i\Pi_k \ \cdots \ \text{cyclically}, \\ Tr\Pi_0 = 4, \\ Tr\Pi_i = 0. \end{array}\right\} \tag{3.34}$$

These results are identical to the relations (2.16). ∎

Since $\Lambda_\mu \in \mathcal{L}$ and $\Pi_\mu \in \mathcal{R}$, the lambda matrices commute with the pi matrices:

$$[\Lambda_\mu, \Pi_\nu] = 0. \tag{3.35}$$

This result is intuitively evident if we compare the constructions of the matrices lambda and pi:

— The lambda matrices are obtained by substituting the sigma matrices for the numerical units in the unit 2×2 matrix. For example,

$$I = \begin{pmatrix} 1 & 0 \\ 0 & 1 \end{pmatrix} \rightarrow \Lambda_2 = \begin{pmatrix} \sigma_2 & 0 \\ 0 & \sigma_2 \end{pmatrix}.$$

— The pi matrices are obtained by substituting the unit 2×2 matrix for the numerical units in the sigma matrices. For example,

$$\sigma_2 = \begin{pmatrix} 0 & -i \\ i & 0 \end{pmatrix} \rightarrow \Pi_2 = \begin{pmatrix} 0 & -iI \\ iI & 0 \end{pmatrix}.$$

Chapter 4

The Algebraic Bases in $\mathcal{M}$

The algebraic properties of the left and right algebras of quantions and of their product are most succinctly and transparently represented by a system of basis matrices in the totality $\mathcal{M}$ of complex 4×4 matrices. A convenient formalism for such a basis is developed in the present chapter.

As a complex linear space, $\mathcal{M}$ is spanned by any system of sixteen linearly independent matrices. But since $\mathcal{M}$ is also an algebra, the presence of an algebraic product introduces a structuring of the basis matrices by way of generators.

In any matrix algebra, the trace distinguishes a particular matrix: the unit matrix I. Thus, the algebra $\mathcal{M}$ naturally decomposes into two parts: the one-dimensional subspace $\mathbb{C}I$ (the center of the algebra), and the complementary 15-dimensional subspace $\mathcal{M}_0$ of all traceless matrices.

By the definition of a basis, the sixteen basis matrices in $\mathcal{M}$ must be linearly independent, but they cannot be algebraically independent. A minimal system of algebraically independent basis matrices in $\mathcal{M}$ is referred to as a system of generators. Since $\dim(\mathcal{M}) = 16$, there are four generators.

In selecting the generators of an algebra, and hence the hierarchy of basis matrices, one tries to bring out the structures best suited to the envisaged applications. Two different bases are thus defined in the algebra $\mathcal{M}$: The well-known basis generated by Dirac's gamma matrices, and a new basis generated by the lambda matrices.

4.1   The Dirac basis

Dirac's gamma matrices γ^μ define a system of basis matrices in $\mathcal{M}$ (the Von Neumann's matrices Γ) which is structured as follows:

$$
\begin{array}{rcl}
\text{16 basis elements} & = & \text{1 unit matrix, } I. \\
& + & \text{4 traceless generators, } \gamma^\mu. \\
& + & \text{11 traceless products of generators}
\end{array}
\tag{4.1}
$$

The complete system of gamma matrices in the chiral representation (including γ^5) is listed below for reference:

$$
\gamma^0 = \begin{pmatrix} 0 & \sigma_0 \\ \sigma_0 & 0 \end{pmatrix} = \begin{pmatrix} 0 & 0 & 1 & 0 \\ 0 & 0 & 0 & 1 \\ 1 & 0 & 0 & 0 \\ 0 & 1 & 0 & 0 \end{pmatrix},
\tag{4.2}
$$

$$
\gamma^1 = \begin{pmatrix} 0 & \sigma_1 \\ -\sigma_1 & 0 \end{pmatrix} = \begin{pmatrix} 0 & 0 & 0 & 1 \\ 0 & 0 & 1 & 0 \\ 0 & -1 & 0 & 0 \\ -1 & 0 & 0 & 0 \end{pmatrix},
\tag{4.3}
$$

$$
\gamma^2 = \begin{pmatrix} 0 & \sigma_2 \\ -\sigma_2 & 0 \end{pmatrix} = \begin{pmatrix} 0 & 0 & 0 & -i \\ 0 & 0 & i & 0 \\ 0 & i & 0 & 0 \\ -i & 0 & 0 & 0 \end{pmatrix},
\tag{4.4}
$$

$$
\gamma^3 = \begin{pmatrix} 0 & \sigma_3 \\ -\sigma_3 & 0 \end{pmatrix} = \begin{pmatrix} 0 & 0 & 1 & 0 \\ 0 & 0 & 0 & -1 \\ -1 & 0 & 0 & 0 \\ 0 & 1 & 0 & 0 \end{pmatrix},
\tag{4.5}
$$

$$
\gamma^5 \overset{def}{=} i\gamma^0\gamma^1\gamma^2\gamma^3 = \begin{pmatrix} -\sigma_0 & 0 \\ 0 & \sigma_0 \end{pmatrix} = \begin{pmatrix} -1 & 0 & 0 & 0 \\ 0 & -1 & 0 & 0 \\ 0 & 0 & 1 & 0 \\ 0 & 0 & 0 & 1 \end{pmatrix}.
\tag{4.6}
$$

In addition to these five matrices with standard symbols γ^0 to γ^5, a sixth antisymmetric matrix occurs in Part II. We shall denote it by

γ^6. It is defined as:

$$\gamma^6 = i\gamma^1\gamma^2\gamma^3 = \gamma^0\gamma^5$$

$$= \begin{pmatrix} 0 & \sigma_0 \\ -\sigma_0 & 0 \end{pmatrix} = \begin{pmatrix} 0 & 0 & 1 & 0 \\ 0 & 0 & 0 & 1 \\ -1 & 0 & 0 & 0 \\ 0 & -1 & 0 & 0 \end{pmatrix}. \tag{4.7}$$

The standard lattice structuring of $\mathcal{M}$ based on the gamma matrices is shown below in the sequence (4.8). The notations $\gamma^{[\cdots]}$ stand for complete anti-symmetrization. For example:

$$\gamma^{[\mu\nu\rho]} \equiv \gamma^{[\mu}\gamma^\nu\gamma^{\rho]}$$

$$= \frac{1}{3!}\left(\gamma^\mu\gamma^\nu\gamma^\rho + \gamma^\rho\gamma^\mu\gamma^\nu + \gamma^\nu\gamma^\rho\gamma^\mu - \gamma^\nu\gamma^\mu\gamma^\rho - \gamma^\mu\gamma^\rho\gamma^\nu - \gamma^\rho\gamma^\nu\gamma^\mu\right).$$

While various other notations can be found in the literature, the symbols $\gamma^{[\cdots]}$ have the advantage of being self-defining (we note that the products of different gamma matrices are automatically antisymmetric, which simplifies calculations):

$$
\begin{array}{rll}
\text{Scalar:} & I & \text{1 dim} \\
& \downarrow & \\
\text{Vector:} & \gamma^\mu & \text{4 dim} \\
& \downarrow & \\
\text{Bivector:} & i\gamma^{[\mu\nu]} & \text{6 dim} \\
& \downarrow & \\
\text{Pseudo vector:} & \gamma^{[\mu\nu\rho]} & \text{4 dim} \\
& \downarrow & \\
\text{Pseudo scalar:} & i\gamma^{[\mu\nu\rho\sigma]} & \text{1 dim}
\end{array}
\tag{4.8}
$$

The fifteen matrices $\gamma^{[\cdots]}$ are explicitly listed for reference in the following table:

The Dirac matrices	
$\gamma^0 = \begin{pmatrix} 0 & \sigma_0 \\ \sigma_0 & 0 \end{pmatrix}$	$\gamma^1 = \begin{pmatrix} 0 & \sigma_1 \\ -\sigma_1 & 0 \end{pmatrix}$
$\gamma^2 = \begin{pmatrix} 0 & \sigma_2 \\ -\sigma_2 & 0 \end{pmatrix}$	$\gamma^3 = \begin{pmatrix} 0 & \sigma_3 \\ -\sigma_3 & 0 \end{pmatrix}$
The bivector	
$\gamma^{[01]} = - \begin{pmatrix} \sigma_1 & 0 \\ 0 & -\sigma_1 \end{pmatrix}$	$\gamma^{[12]} = -i \begin{pmatrix} \sigma_3 & 0 \\ 0 & \sigma_3 \end{pmatrix}$
$\gamma^{[02]} = - \begin{pmatrix} \sigma_2 & 0 \\ 0 & -\sigma_2 \end{pmatrix}$	$\gamma^{[23]} = -i \begin{pmatrix} \sigma_1 & 0 \\ 0 & \sigma_1 \end{pmatrix}$
$\gamma^{[03]} = - \begin{pmatrix} \sigma_3 & 0 \\ 0 & -\sigma_3 \end{pmatrix}$	$\gamma^{[31]} = -i \begin{pmatrix} \sigma_2 & 0 \\ 0 & \sigma_2 \end{pmatrix}$
The pseudo vector	
$\gamma^{[123]} = -i \begin{pmatrix} 0 & \sigma_0 \\ -\sigma_0 & 0 \end{pmatrix}$	$\gamma^{[023]} = -i \begin{pmatrix} 0 & \sigma_1 \\ \sigma_1 & 0 \end{pmatrix}$
$\gamma^{[031]} = -i \begin{pmatrix} 0 & \sigma_2 \\ \sigma_2 & 0 \end{pmatrix}$	$\gamma^{[012]} = -i \begin{pmatrix} 0 & \sigma_3 \\ \sigma_3 & 0 \end{pmatrix}$
The pseudo scalar	
$i\gamma^{[0123]} = \begin{pmatrix} -\sigma_0 & 0 \\ 0 & \sigma_0 \end{pmatrix} = \gamma^5$	

Table 4.1: The Dirac basis in $\mathcal{M}$.

4.2   The quantionic basis

The lambda and pi matrices defined in Sections 2.2 and 3.5 respectively provide seven basis matrices in $\mathcal{M}$. Thus, nine more are needed to complete the set. They are most naturally provided by the nine mixed products of lambda and pi matrices.

The notation $\Omega_{\mu\nu}$ will be used for the sixteen different products of the basis matrices Λ_μ and Π_ν :

$$\Omega_{\mu\nu} \overset{def}{=} \Lambda_\mu \Pi_\nu. \tag{4.9}$$

The special cases are

$$\left.\begin{aligned}
\Omega_{00} &= I, \\
\Omega_{i0} &= \Lambda_i \\
\Omega_{0j} &= \Pi_j, \\
\Omega_{ij} &= \Lambda_i\Pi_j,
\end{aligned}\right\} \tag{4.10}$$

To the tensorial structuring (4.8) of the Dirac basis corresponds the multiplicative structuring of the quantionic basis shown in the following table :

	Π_0	Π_1	Π_2	Π_3
Λ_0	I	Π_1	Π_2	Π_3
Λ_1	Λ_1	Ω_{11}	Ω_{12}	Ω_{13}
Λ_2	Λ_2	Ω_{21}	Ω_{22}	Ω_{23}
Λ_3	Λ_3	Ω_{31}	Ω_{32}	Ω_{33}

The bold-framed matrices Λ_1, Λ_2, Π_1, Π_2 are possible choices of generators because

$$\left.\begin{aligned}
\Lambda_1\Lambda_2 &= i\Lambda_3, \\
\Pi_1\Pi_2 &= i\Pi_3.
\end{aligned}\right\} \tag{4.11}$$

Theorem 17 *The matrices $\Omega_{\mu\nu}$ form a basis in $\mathcal{M}$.*

Proof. We only have to show that the nine matrices Ω_{ij} are different. To this end, we solve the equation $\Omega_{ij} = \Omega_{mn}$, that is,

$$\Lambda_i \Pi_j = \Lambda_m \Pi_n.$$

Multiplying both sides from the left by Λ_i and from the right by Π_n, one obtains

$$\Pi_j \Pi_n = \Lambda_i \Lambda_m,$$

which is true if and only if $j = n$ and $i = m$. ∎

The matrices $\Omega_{\mu\nu}$ are displayed for reference in the following table:

Basis in the traceless part of $\mathcal{L}$		
$\Lambda_1 = \begin{pmatrix} \sigma_1 & 0 \\ 0 & \sigma_1 \end{pmatrix}$	$\Lambda_2 = \begin{pmatrix} \sigma_2 & 0 \\ 0 & \sigma_2 \end{pmatrix}$	$\Lambda_3 = \begin{pmatrix} \sigma_3 & 0 \\ 0 & \sigma_3 \end{pmatrix}$
Basis in the traceless part of $\mathcal{R}$		
$\Pi_1 = \begin{pmatrix} 0 & \sigma_0 \\ \sigma_0 & 0 \end{pmatrix}$	$\Pi_2 = -i \begin{pmatrix} 0 & \sigma_0 \\ -\sigma_0 & 0 \end{pmatrix}$	$\Pi_3 = \begin{pmatrix} \sigma_0 & 0 \\ 0 & -\sigma_0 \end{pmatrix}$
Basis in the traceless part of $\mathcal{M} \setminus (\mathcal{L} \cup \mathcal{R})$		
$\Omega_{11} = \begin{pmatrix} 0 & \sigma_1 \\ \sigma_1 & 0 \end{pmatrix}$	$\Omega_{12} = -i \begin{pmatrix} 0 & \sigma_1 \\ -\sigma_1 & 0 \end{pmatrix}$	$\Omega_{13} = \begin{pmatrix} \sigma_1 & 0 \\ 0 & -\sigma_1 \end{pmatrix}$
$\Omega_{21} = \begin{pmatrix} 0 & \sigma_2 \\ \sigma_2 & 0 \end{pmatrix}$	$\Omega_{22} = -i \begin{pmatrix} 0 & \sigma_2 \\ -\sigma_2 & 0 \end{pmatrix}$	$\Omega_{23} = \begin{pmatrix} \sigma_2 & 0 \\ 0 & -\sigma_2 \end{pmatrix}$
$\Omega_{31} = \begin{pmatrix} 0 & \sigma_3 \\ \sigma_3 & 0 \end{pmatrix}$	$\Omega_{32} = -i \begin{pmatrix} 0 & \sigma_3 \\ -\sigma_3 & 0 \end{pmatrix}$	$\Omega_{33} = \begin{pmatrix} \sigma_3 & 0 \\ 0 & -\sigma_3 \end{pmatrix}$

Table 4.2: The quantionic basis in the traceless part of $\mathcal{M}$.

Comparison of the Tables 4.1 and 4.2 yields the following one-to-one mappings between the Dirac basis matrices $\gamma^{[\cdots]}$ and the quantionic basis matrices $\Omega_{\mu\nu}$:

Basis in $\mathcal{L}$		
$\Lambda_1 = i\gamma^{[23]}$	$\Lambda_2 = i\gamma^{[31]}$	$\Lambda_3 = i\gamma^{[12]}$
Basis in $\mathcal{R}$		
$\Pi_1 = \gamma^0$	$\Pi_2 = i\gamma^{[123]}$	$\Pi_3 = -i\gamma^{[0123]}$
	$= -i\gamma^6$	$= -\gamma^5$
Basis in $\mathcal{M} \setminus (\mathcal{L} \cup \mathcal{R})$		
$\Omega_{11} = i\gamma^{[023]}$	$\Omega_{12} = -i\gamma^1$	$\Omega_{13} = -\gamma^{[01]}$
$\Omega_{21} = i\gamma^{[031]}$	$\Omega_{22} = -i\gamma^2$	$\Omega_{23} = -\gamma^{[02]}$
$\Omega_{31} = i\gamma^{[012]}$	$\Omega_{32} = -i\gamma^3$	$\Omega_{33} = -\gamma^{[03]}$

$$(4.12)$$

While patterns can be recognized within the columns, these relations are far from obvious. The reason is that the two set of basis matrices are structured by very different principles: The set of matrices $\gamma^{[\cdots]}$ is structured as a Grassmann algebra; the set of matrices $\Omega_{\mu\nu}$ is structured as a 4×4 multiplication table.

Theorem 18 *The omega basis matrices satisfy the following symmetry relations:*

$$\Lambda_\mu \Pi_\nu = \Pi_\nu \Lambda_\mu, \qquad (4.13)$$
$$\Omega_{\mu\nu} = W \Omega_{\nu\mu} W, \qquad (4.14)$$
$$\Omega_{\mu\nu}^\dagger = \Omega_{\mu\nu}, \qquad (4.15)$$

Proof. Since $\Lambda_\mu \in \mathcal{L}$ and $\Pi_\nu \in \mathcal{R}$, the commutation relation (4.13) is an immediate consequence of the fact that the subalgebras $\mathcal{L}$ and $\mathcal{R}$ are mutual commutants.

The proof of relation (4.14) consists of the following sequence of equations (note that $WW = I$):

$$
\begin{aligned}
W\Omega_{\nu\mu}W &= W\Lambda_\nu\Pi_\mu W = W\Lambda_\nu\,(WW)\,\Pi_\mu W \\
&= (W\Lambda_\nu W)\,(W\Pi_\mu W) = \Pi_\nu\Lambda_\mu.
\end{aligned}
$$

Hence, by (4.13):

$$
W\Omega_{\nu\mu}W = \Pi_\nu\Lambda_\mu = \Lambda_\mu\Pi_\nu = \Omega_{\mu\nu}.
$$

Relation (4.15) follows from the sequence of identities

$$
\begin{aligned}
\Omega_{\mu\nu}^\dagger &= (\Lambda_\mu\Pi_\nu)^\dagger = \Pi_\nu^\dagger\Lambda_\mu^\dagger = \Pi_\nu\Lambda_\mu = \Lambda_\mu\Pi_\nu \\
&= \Omega_{\mu\nu}.
\end{aligned}
$$

Thus, the basis matrices $\Omega_{\mu\nu}$ are Hermitian. ∎

The defining relations for the Pauli matrices,

$$
\sigma_i\sigma_j = i\varepsilon_{ijk}\sigma_k + \delta_{ij}I,
$$

extend to the lambda and pi matrices:

$$
\left.
\begin{aligned}
\Lambda_i\Lambda_j &= i\varepsilon_{ijk}\Lambda_k + \delta_{ij}I, \\
\Pi_i\Pi_j &= i\varepsilon_{ijk}\Pi_k + \delta_{ij}I.
\end{aligned}
\right\}
\tag{4.16}
$$

Taking into account the commutativity of $\mathcal{L}$ and $\mathcal{R}$, which implies

$$
\Lambda_i\Pi_j = \Pi_j\Lambda_i = \Omega_{ij},
$$

one obtains the multiplication rules for the matrices $\Omega_{\mu\nu}$:

$$
\Omega_{\mu\nu}\Omega_{\rho\tau} = \Lambda_\mu\Pi_\nu\Lambda_\rho\Pi_\tau = \Lambda_\mu\Lambda_\rho\Pi_\nu\Pi_\tau.
\tag{4.17}
$$

These expressions can be explicitly computed from the relations (4.16) or from the Table 3.2. The trace part of the product,

$$
\Omega_{\mu\nu}\Omega_{\rho\tau} = \delta_{\mu\rho}\delta_{\nu\tau}I + (\text{a traceless matrix}),
\tag{4.18}
$$

is the only one we shall need.

4.3 Tabulated products of basis elements

The products of omega matrices

	Λ_1	Λ_2	Λ_3	Π_1	Π_2	Π_3
Λ_1	I	$i\Lambda_3$	$-i\Lambda_2$	Ω_{11}	Ω_{12}	Ω_{13}
Λ_2	$-i\Lambda_3$	I	$i\Lambda_1$	Ω_{21}	Ω_{22}	Ω_{23}
Λ_3	Λ_2	$-i\Lambda_1$	I	Ω_{31}	Ω_{32}	Ω_{33}
Π_1	Ω_{11}	Ω_{21}	Ω_{31}	I	$i\Pi_3$	$-i\Pi_2$
Π_2	Ω_{12}	Ω_{22}	Ω_{32}	$-i\Pi_3$	I	$i\Pi_1$
Π_3	Ω_{13}	Ω_{23}	Ω_{33}	$i\Pi_2$	$i\Pi_1$	I
Ω_{11}	Π_1	$i\Omega_{31}$	$-i\Omega_{21}$	Λ_1	$i\Omega_{13}$	$-i\Omega_{12}$
Ω_{12}	Π_2	$i\Omega_{32}$	$-i\Omega_{22}$	$-i\Omega_{13}$	Λ_1	$i\Omega_{11}$
Ω_{13}	Π_3	$i\Omega_{33}$	$-i\Omega_{23}$	$i\Omega_{12}$	$-i\Omega_{11}$	Λ_1
Ω_{21}	$-i\Omega_{31}$	Π_1	$i\Omega_{11}$	Λ_2	$i\Omega_{23}$	$-i\Omega_{22}$
Ω_{22}	$-i\Omega_{32}$	Π_2	$i\Omega_{12}$	$-i\Omega_{23}$	Λ_2	$i\Omega_{21}$
Ω_{23}	$-i\Omega_{33}$	Π_3	$i\Omega_{13}$	$i\Omega_{22}$	$-i\Omega_{21}$	Λ_2
Ω_{31}	$i\Omega_{21}$	$-i\Omega_{11}$	Π_1	Λ_3	$i\Omega_{33}$	$-i\Omega_{32}$
Ω_{32}	$i\Omega_{22}$	$-i\Omega_{12}$	Π_2	$-i\Omega_{33}$	Λ_3	$i\Omega_{31}$
Ω_{33}	$i\Omega_{23}$	$-i\Omega_{13}$	Π_3	$i\Omega_{32}$	$-i\Omega_{31}$	Λ_3

The products of omega matrices (continued)

	Ω_{11}	Ω_{12}	Ω_{13}	Ω_{21}	Ω_{22}	Ω_{23}	Ω_{31}	Ω_{32}	Ω_{33}
Λ_1	Π_1	Π_2	Π_3	$i\Omega_{31}$	$i\Omega_{32}$	$i\Omega_{33}$	$-i\Omega_{21}$	$-i\Omega_{22}$	$-i\Omega_{23}$
Λ_2	$-i\Omega_{31}$	$-i\Omega_{32}$	$-i\Omega_{33}$	Π_1	Π_2	Π_3	$i\Omega_{11}$	$i\Omega_{12}$	$i\Omega_{13}$
Λ_3	$i\Omega_{21}$	$i\Omega_{22}$	$i\Omega_{23}$	$-i\Omega_{31}$	$-i\Omega_{32}$	$-i\Omega_{33}$	Π_1	Π_2	Π_3
Π_1	Λ_1	$i\Omega_{13}$	$-i\Omega_{12}$	Λ_2	$i\Omega_{23}$	$-i\Omega_{22}$	Λ_3	$i\Omega_{33}$	$-i\Omega_{32}$
Π_2	$-1\Omega_{13}$	Λ_1	$i\Omega_{11}$	$-i\Omega_{23}$	Λ_2	$i\Omega_{21}$	$-i\Omega_{33}$	Λ_3	$i\Omega_{31}$
Π_3	$i\Omega_{12}$	$-i\Omega_{11}$	Λ_1	$i\Omega_{22}$	$-i\Omega_{21}$	Λ_2	$i\Omega_{32}$	$-i\Omega_{31}$	Λ_3
Ω_{11}	I	$i\Pi_3$	$-i\Pi_2$	$i\Lambda_3$	$-\Omega_{33}$	Ω_{22}	$-i\Lambda_2$	Ω_{23}	$-\Omega_{22}$
Ω_{12}	$-i\Pi_3$	I	$i\Pi_1$	Ω_{33}	$i\Lambda_3$	$-\Omega_{31}$	$-\Omega_{23}$	$-i\Lambda_2$	Ω_{21}
Ω_{13}	$i\Pi_2$	$-i\Pi_1$	I	$-\Omega_{32}$	Ω_{31}	$i\Lambda_3$	Ω_{22}	$-\Omega_{21}$	$-i\Lambda_2$
Ω_{21}	$-i\Lambda_3$	Ω_{33}	$-\Omega_{32}$	I	$i\Pi_3$	$-i\Pi_2$	$i\Lambda_1$	$-\Omega_{13}$	Ω_{12}
Ω_{22}	$-\Omega_{33}$	$-i\Lambda_3$	Ω_{31}	$-i\Pi_3$	I	$i\Pi_1$	Ω_{13}	$i\Lambda_1$	$-\Omega_{11}$
Ω_{23}	Ω_{32}	$-\Omega_{31}$	$-i\Lambda_3$	$i\Pi_2$	$-i\Pi_1$	I	$-\Omega_{12}$	$i\Omega_{11}$	$i\Lambda_1$
Ω_{31}	$i\Lambda_2$	$-\Omega_{23}$	Ω_{22}	$-i\Lambda_1$	Ω_{13}	$-\Omega_{12}$	I	$i\Pi_3$	$-i\Pi_2$
Ω_{32}	Ω_{23}	$i\Lambda_2$	$-\Omega_{21}$	$-\Omega_{13}$	$-i\Lambda_1$	Ω_{11}	$-i\Pi_3$	I	$i\Pi_1$
Ω_{33}	$-\Omega_{22}$	$-\Omega_{31}$	$i\Lambda_2$	Ω_{12}	$-i\Omega_{11}$	$-i\Lambda_1$	$i\Pi_2$	$-i\Pi_1$	I

The elementary basis matrices in $\mathcal{M}$

$$\Delta_{11} = \begin{pmatrix} 1 & 0 & 0 & 0 \\ 0 & 0 & 0 & 0 \\ 0 & 0 & 0 & 0 \\ 0 & 0 & 0 & 0 \end{pmatrix} = \tfrac{1}{4}\left(I + \Pi_3 + \Lambda_3 + \Omega_{33}\right)$$

$$\Delta_{12} = \begin{pmatrix} 0 & 1 & 0 & 0 \\ 0 & 0 & 0 & 0 \\ 0 & 0 & 0 & 0 \\ 0 & 0 & 0 & 0 \end{pmatrix} = \tfrac{1}{4}\left(\Lambda_1 + \Omega_{13} + i\Lambda_2 + i\Omega_{23}\right)$$

$$\Delta_{13} = \begin{pmatrix} 0 & 0 & 1 & 0 \\ 0 & 0 & 0 & 0 \\ 0 & 0 & 0 & 0 \\ 0 & 0 & 0 & 0 \end{pmatrix} = \tfrac{1}{4}\left(\Pi_1 + i\Pi_2 + \Omega_{31} + i\Omega_{32}\right)$$

$$\Delta_{14} = \begin{pmatrix} 0 & 0 & 0 & 1 \\ 0 & 0 & 0 & 0 \\ 0 & 0 & 0 & 0 \\ 0 & 0 & 0 & 0 \end{pmatrix} = \tfrac{1}{4}\left(\Omega_{11} - \Omega_{22} + i\Omega_{12} + i\Omega_{21}\right)$$

$$\Delta_{21} = \begin{pmatrix} 0 & 0 & 0 & 0 \\ 1 & 0 & 0 & 0 \\ 0 & 0 & 0 & 0 \\ 0 & 0 & 0 & 0 \end{pmatrix} = \tfrac{1}{4}\left(\Lambda_1 + \Omega_{13} - i\Lambda_2 - i\Omega_{23}\right)$$

$$\Delta_{22} = \begin{pmatrix} 0 & 0 & 0 & 0 \\ 0 & 1 & 0 & 0 \\ 0 & 0 & 0 & 0 \\ 0 & 0 & 0 & 0 \end{pmatrix} = \tfrac{1}{4}\left(I + \Pi_3 - \Lambda_3 - \Omega_{33}\right)$$

$$\Delta_{23} = \begin{pmatrix} 0 & 0 & 0 & 0 \\ 0 & 0 & 1 & 0 \\ 0 & 0 & 0 & 0 \\ 0 & 0 & 0 & 0 \end{pmatrix} = \tfrac{1}{4}\left(\Omega_{11} + i\Omega_{12} + \Omega_{22} - i\Omega_{21}\right)$$

$$\Delta_{24} = \begin{pmatrix} 0 & 0 & 0 & 0 \\ 0 & 0 & 0 & 1 \\ 0 & 0 & 0 & 0 \\ 0 & 0 & 0 & 0 \end{pmatrix} = \tfrac{1}{4}\left(\Pi_1 + i\Pi_2 - \Omega_{31} - i\Omega_{32}\right)$$

The elementary basis matrices in $\mathcal{M}$ (continued)

Δ_{31} =	$\begin{pmatrix} 0 & 0 & 0 & 0 \\ 0 & 0 & 0 & 0 \\ 1 & 0 & 0 & 0 \\ 0 & 0 & 0 & 0 \end{pmatrix}$	= $\frac{1}{4}\left(\Pi_1 - i\Pi_2 + \Omega_{31} - i\Omega_{32}\right)$
Δ_{32} =	$\begin{pmatrix} 0 & 0 & 0 & 0 \\ 0 & 0 & 0 & 0 \\ 0 & 1 & 0 & 0 \\ 0 & 0 & 0 & 0 \end{pmatrix}$	= $\frac{1}{4}\left(\Omega_{11} + \Omega_{22} + i\Omega_{21} - i\Omega_{12}\right)$
Δ_{33} =	$\begin{pmatrix} 0 & 0 & 0 & 0 \\ 0 & 0 & 0 & 0 \\ 0 & 0 & 1 & 0 \\ 0 & 0 & 0 & 0 \end{pmatrix}$	= $\frac{1}{4}\left(I - \Pi_3 + \Lambda_3 - \Omega_{33}\right)$
Δ_{34} =	$\begin{pmatrix} 0 & 0 & 0 & 0 \\ 0 & 0 & 0 & 0 \\ 0 & 0 & 0 & 1 \\ 0 & 0 & 0 & 0 \end{pmatrix}$	= $\frac{1}{4}\left(\Lambda_1 - \Omega_{13} + i\Lambda_2 - i\Omega_{23}\right)$
Δ_{41} =	$\begin{pmatrix} 0 & 0 & 0 & 0 \\ 0 & 0 & 0 & 0 \\ 0 & 0 & 0 & 0 \\ 1 & 0 & 0 & 0 \end{pmatrix}$	= $\frac{1}{4}\left(\Omega_{11} - i\Omega_{12} - i\Omega_{21} - \Omega_{22}\right)$
Δ_{42} =	$\begin{pmatrix} 0 & 0 & 0 & 0 \\ 0 & 0 & 0 & 0 \\ 0 & 0 & 0 & 0 \\ 0 & 1 & 0 & 0 \end{pmatrix}$	= $\frac{1}{4}\left(\Pi_1 - i\Pi_2 - \Omega_{31} + i\Omega_{32}\right)$
Δ_{43} =	$\begin{pmatrix} 0 & 0 & 0 & 0 \\ 0 & 0 & 0 & 0 \\ 0 & 0 & 0 & 0 \\ 0 & 0 & 1 & 0 \end{pmatrix}$	= $\frac{1}{4}\left(\Lambda_1 - \Omega_{13} - i\Lambda_2 + i\Omega_{23}\right)$
Δ_{44} =	$\begin{pmatrix} 0 & 0 & 0 & 0 \\ 0 & 0 & 0 & 0 \\ 0 & 0 & 0 & 0 \\ 0 & 0 & 0 & 1 \end{pmatrix}$	= $\frac{1}{4}\left(I - \Pi_3 - \Lambda_3 + \Omega_{33}\right)$

4.4   The decomposition of Hermitian matrices

Any maximal system of linearly independent matrices forms a basis in the complex matrix algebra $\mathcal{M}$, but if the basis matrices are Hermitian (which is the case for the quantionic basis but not for the Dirac basis), two substantial simplifications take place:

(a) A matrix $M \in \mathcal{M}$ is Hermitian if and only if the coefficients of its expansion are real.

(b) Hermitian conjugation is equivalent to the complex conjugation of the coefficients.

It is thus most natural and most elegant to select in $\mathcal{M}$ a basis which is Hermitian, provided some other condition that takes precedence is not inconsistent with such a choice.

In quantum field theory, Dirac's objective was to take the square root of the D'Alambert operator:

$$\Box = \eta^{\mu\nu}\partial_\mu\partial_\nu = (\gamma^\mu\partial_\mu)^2. \tag{4.19}$$

Consequently, this particular factorization condition takes precedence over Hermiticity. It defines Dirac's derivation operator $\gamma^\mu\partial_\mu$, and implies that the matrices γ^μ must satisfy the equation

$$\gamma^\mu\gamma^\nu + \gamma^\nu\gamma^\mu = 2\eta^{\mu\nu}I. \tag{4.20}$$

The solutions for the gamma matrices are not Hermitian. To see why, take $\mu = \nu = i$ (where i is 1, 2, or 3). The condition (4.20) reduces to the special case $(\gamma^i)^2 = \eta^{ii}I = -I$, and since the trace of the square of any Hermitian matrix is positive-definite, it follows that the matrices γ^i are antihermitian.

On the other hand, there exists in the algebra $\mathcal{M}$ a Hermitian basis in which the D'Alambert operator admits a factorization other than a square:

$$\Box = \eta^{\mu\nu}\partial_\mu\partial_\nu = \mathcal{D}^\#\mathcal{D}. \tag{4.21}$$

We shall see later that the derivation operator $\mathcal{D}$ is the quantionic version of the Dirac operator $\gamma^\mu\partial_\mu$. As a matrix, the operator $\mathcal{D}$ is Hermitian. The operator $\mathcal{D}^\#$ is referred to as the "metric dual" of $\mathcal{D}$, a concept defined in Chapter 6. The Hermitian basis in which the factorization (4.21) is possible is the system of matrices $\Omega_{\mu\nu}$.

If $Z \in \mathcal{M}$ is an arbitrary 4×4 complex matrix, it's expansion in the basis $(\Omega_{\mu\nu})$ may be written in the form

$$Z = z^{\mu\nu}\Omega_{\mu\nu}. \tag{4.22}$$

Let us make several observations to preclude misinterpretations.

(1) The sixteen coefficients $z^{\mu\nu}$ are complex numbers. In the matrix $(z^{\mu\nu})$ of these coefficients, the indices μ and ν specify matrix elements, not tensor indices.

(2) The sixteen objects $\Omega_{\mu\nu}$ are matrices. Thus, the indices μ and ν do not specify matrix elements. They *label* matrices according to the definitions listed in Table 4.12 on page 61.

(3) The summation convention assumed in relation (4.22) is strictly for convenience. While borrowed from the formalism of tensor algebra, the objects themselves are not regarded as tensors — at least not in the present work.

Given a quantion Z, one obtains the expansion coefficients $z^{\alpha\beta}$ in relation (4.22) by taking the trace of the product $Z\Omega_{\alpha\beta}$:

$$z^{\alpha\beta} = \frac{1}{2}\left(\omega|Z\Omega_{\alpha\beta}|\omega\right). \tag{4.23}$$

The coefficient is $1/2$ because $z^{\alpha\beta}$ is $1/4$ of the trace, and, by relation (2.28), the trace is twice $(\omega| * |\omega)$.

The basis vectors $\Omega^{\alpha\beta}$ being Hermitian, a matrix H is Hermitian if and only if the coefficients $h^{\mu\nu}$ in the expansion

$$H = h^{\mu\nu}\Omega_{\mu\nu} \tag{4.24}$$

are real numbers.

As an immediate application of the rule (4.23), we might expand the distinguished matrix W given by relation (3.13),

$$z^{\alpha\beta} = \frac{1}{2}\left(\omega|W\Omega_{\alpha\beta}|\omega\right),$$

but this involves computing 16 coefficients. Since 12 of them vanish, it is practically much simpler the use the table of elementary bases displayed on page 65,

$$W = \begin{pmatrix} 1 & 0 & 0 & 0 \\ 0 & 0 & 1 & 0 \\ 0 & 1 & 0 & 0 \\ 0 & 0 & 0 & 1 \end{pmatrix} = \Delta_{11} + \Delta_{23} + \Delta_{32} + \Delta_{44}.$$

This yields:

$$W = \frac{1}{2} \left(I + \Omega_{11} + \Omega_{22} + \Omega_{33} \right). \tag{4.25}$$

Since a general complex matrix M is a unique linear combination of two Hermitian matrices, $M = H_1 + iH_2$, it suffices to derive theorems about linear properties of matrices for Hermitian matrices only. The following is such a theorem:

Theorem 19 *An arbitrary Hermitian matrix $H \in \mathcal{M}^h$ admits two decompositions. One decomposition is in terms of coefficients L^0 to L^3 that are Hermitian L-type quantions:*

$$\left. \begin{array}{l} H = L^0 \Pi_0 + L^1 \Pi_1 + L^2 \Pi_2 + L^3 \Pi_3, \\[2mm] L^0, L^1, L^2, L^3 \in \mathcal{L}^h. \end{array} \right\} \tag{4.26}$$

The other decomposition is in terms of coefficients R^0 to R^3 that are Hermitian R-type quantions:

$$\left. \begin{array}{l} H = R^0 \Lambda_0 + R^1 \Lambda_1 + R^2 \Lambda_2 + R^3 \Lambda_3, \\[2mm] R^0, R^1, R^2, R^3 \in \mathcal{R}^h. \end{array} \right\} \tag{4.27}$$

Proof. Substitution of the definition (4.9) for $\Omega_{\mu\nu}$ into relation (4.24) written in the form

$$H = H = \Lambda_\mu h^{\mu\nu} \Pi_\nu$$

yields two expressions for H, depending on how we associate the factors:

$$\left(\Lambda_\mu h^{\mu\nu} \right) \Pi_\nu = H = \Lambda_\mu \left(h^{\mu\nu} \Pi_\nu \right).$$

These expressions are identical in content, but they have different interpretations:

(1) On the left-hand side, the term $(\Lambda_\mu h^{\mu\nu})$ represents four Hermitian L-type quantions, or **L-type coefficients**:

$$L^\nu = h^{\mu\nu}\Lambda_\mu \in \mathcal{L}^h.$$

(2) On the right-hand side, the term $(h^{\mu\nu}\Pi_\nu)$ represents four Hermitian R-type quantions, or **R-type coefficients**:

$$R^\mu = h^{\mu\nu}\Pi_\nu \in \mathcal{R}^h.$$

This proves the theorem. ∎

We shall conclude this chapter with two intuition-building discussions of some of the insights obtained above.

A discussion of the factorization of the D'Alambertian

Since the D'Alambertian $\Box$ admits the two different factorizations (4.19) and (4.21), a conceptual analysis of the factorization procedure itself might be rather instructive at this point.

To begin with a clean slate, let us go back in thought to a time before Dirac factorized the D'Alambertian, and let us imagine that we are to find such a factorization. Being unbiased by Dirac's solution, by the physically motivated research in Clifford algebras, and by the quantionic approach developed in the present volume — all of which are still in the future — we might have argued as follows:

1. A scalar factorization of the D'Alambertian being impossible, let us seek, for some n that is not known *a priori,* an $n \times n$ differential matrix operator, **D**, whose elements are some homogeneous linear combinations, $a^\mu \partial_\mu$, of the first order partial differential operators in the Minkowski space.

2. Let another such operator, denoted by $\mathbf{D}^\circ$, be uniquely defined by the operator **D** under the condition $\mathbf{D}^{\circ\circ} = \mathbf{D}$.

Statement of the problem: Find all solutions for n, **D**, and $\mathbf{D}^\circ$, that satisfy the operator identity

$$\mathbf{D}^\circ\mathbf{D} = \mathbf{D}\mathbf{D}^\circ = \Box I,$$

where I is the $n \times n$ unit matrix.

Clearly, the assumption $\mathbf{D}^\circ = \mathbf{D}$ yields Dirac's solution.

Historically, once Dirac's factorization solved the problem that led to it (the relativistic generalization of Schrödinger's equation), there was no motivation to seek other solutions characterized by $\mathbf{D}^\circ \neq \mathbf{D}$, or even to suspect that such solutions might exist and be physically relevant.

Yet, if one assumes that $n = 4$ (as suggested by Dirac's solution), it is not very difficult to find a solution for which $\mathbf{D}^\circ \neq \mathbf{D}$, and to prove that it is unique up to similarity transformations. The solution in question is referred to as the "quantionic derivation operator". It is developed in Chapter 13 within the theory of quantions.

The two factorizations (4.19) and (4.21) are closely related by what we call the "quantion-spinor complementarity" — a non-obvious one-to-one correspondence established in Chapter 19 between quantions and Dirac spinors.

A discussion of the decomposition of $\mathcal{M}$

The quantionic structuring of the algebra $\mathcal{M}$ contains some physics which is not present in the standard structuring of the same algebra with the matrices $\gamma^{[\cdots]}$.

We first observe that the systems of basis matrices Λ^μ and Π^μ are analogous to orthonormal systems of basis vectors in geometry, but they are structurally much richer, both mathematically and, as we shall see later, in physical interpretations.

A relevant example is Theorem 19. While mathematically trivial, this theorem plays an essential role in the physical interpretation of quantions. The following brief preview of this role is meant to cast some additional light on the ideas that have been developed in the present chapter.

We shall see in Part II that the Zovko interpretation leads to a unique quantionic field equation by a procedure — referred to as "structural quantization" — which is very closely related to the separation of variables in differential equations. This procedure gives rise to a system of sixteen arbitrary separating functions (separating functions are analogous to integration functions). These functions happen to be structured as a 4×4 Hermitian matrix field $H(x)$ whose physical interpretations are very suggestive:

Mathematically, $H(x)$ is the differential connection which makes differential operators covariant with respect to the quantionic generalization $U_q(1)$ of the gauge group $U(1)$. Physically, the Hermitian matrix $H(x)$ may be interpreted as an external potential. Reminder: The group $U_q(1)$ is isomorphic with the gauge group $U(1) \times SU(2)$ of the electroweak unification.

Let us now consider the decomposition (4.27) of an arbitrary Hermitian matrix H, and let us also decompose the R-type quantions R^μ in the basis of pi matrices.

$$R^\mu = h^{\mu 0}\Pi_0 + h^{\mu 1}\Pi_1 + h^{\mu 2}\Pi_2 + h^{\mu 3}\Pi_3.$$

Dropping the label $\mu = 0, 1, 2, 3$ to reduce clutter, we thus have four R-type quantions, each being of the form

$$R = A^0\Pi_0 + A^1\Pi_1 + A^2\Pi_2 + A^3\Pi_3 = A^\nu\Pi_\nu.$$

It follows that the Hermitian matrix potential H is equivalent to four real vector potentials. This is reminiscent of the four vector potentials A^μ in the electroweak theory.

We shall see in Part II (in the chapters on the derivation of Dirac's equation) that one of the vectors defining the matrix H is indeed the electromagnetic potential. The other three vectors seem to have the correct symmetry properties to be related to the three intermediate vector bosons. At this point, however, this is only a reasonable conjecture because the results obtained so far are semi-classical in the potentials.

Chapter 5

The Complex Numbers and the Quaternions

The algebra of quantions is a unique extension of the field of complex numbers generated by the extension procedure briefly described on page 17. While the procedure itself is not relevant to the objective of this book, the relationship between complex numbers and quantions is essential. This relationship is examined in Section 5.1 by viewing the field of complex numbers as a restriction of the algebra of quantions.

As shown in Table 1.1 on page 14, the other number system in the structural vicinity of the algebra of quantions is the field of quaternions. In Section 5.2, we bring some essential properties of quantions into focus by comparing them with these two structures.

5.1 The restriction of quantions to complex numbers

By Definition 1 on page 27, a general quantion $Q \in \mathcal{L}$ is represented by a matrix of the form

$$Q = \begin{pmatrix} A & 0 \\ 0 & A \end{pmatrix} = \begin{pmatrix} u & w & 0 & 0 \\ v & z & 0 & 0 \\ 0 & 0 & u & w \\ 0 & 0 & v & z \end{pmatrix}, \tag{5.1}$$

where u, v, w, z are arbitrary complex numbers.

Two types of sub-matrices of the matrix Q can be identified with the complex numbers. We shall refer to them as the "algebraic limit" and the "geometric limit". In the former, the imaginary unit is interpreted in the algebraic sense of $i^2 = -1$. In the latter, this unit is interpreted in the geometric sense as a counter-clockwise rotation by $\pi/2$ in the Gaussian plane.

The geometric limit

Multiplication of a complex number by the imaginary unit written in the form $i = e^{i\pi/2}$ represents a counter-clockwise rotation by $\pi/2$ in the Gaussian plane. As a linear operator acting on two-dimensional vectors, such a rotation is represented by the matrix

$$J = \begin{pmatrix} 0 & -1 \\ 1 & 0 \end{pmatrix}. \tag{5.2}$$

Hence, a real linear combination

$$z = xI + yJ = \begin{pmatrix} x & -y \\ y & x \end{pmatrix} \tag{5.3}$$

of the identity mapping I and of the rotation J has the algebraic properties of a complex number. Taking this matrix as the block A in the expression (5.1), one obtains

$$Q = \begin{pmatrix} x & -y & 0 & 0 \\ y & x & 0 & 0 \\ 0 & 0 & x & -y \\ 0 & 0 & y & x \end{pmatrix} \in \mathcal{L}. \tag{5.4}$$

Thus, the subalgebra of matrices of type (5.4) is isomorphic with the field of complex numbers.

The elements of the matrix (5.4) are real, so that the Hermitian conjugate is the transpose:

$$Q^\dagger = \begin{pmatrix} x & y & 0 & 0 \\ -y & x & 0 & 0 \\ 0 & 0 & x & y \\ 0 & 0 & -y & x \end{pmatrix}. \tag{5.5}$$

The algebraic norm of Q, denoted by $A(Q)$, is thus

$$A(Q) = Q^\dagger Q = \begin{pmatrix} x & y & 0 & 0 \\ -y & x & 0 & 0 \\ 0 & 0 & x & y \\ 0 & 0 & -y & x \end{pmatrix} \begin{pmatrix} x & -y & 0 & 0 \\ y & x & 0 & 0 \\ 0 & 0 & x & -y \\ 0 & 0 & y & x \end{pmatrix}$$

$$= (x^2 + y^2) I, \tag{5.6}$$

as expected.

The algebraic limit

Let us consider the "center" of the algebra $\mathcal{L}$, that is, the subalgebra of $\mathcal{L}$ whose elements commute with all elements of $\mathcal{L}$.

Since the sub-matrix A is arbitrary, only multiples of the unit matrix commute with it. It follows that the center of the algebra $\mathcal{L}$ is the subalgebra of matrices

$$Q = zI. \tag{5.7}$$

As stated without proof in Chapter 1, the algebra of quantions is the unique extension of the field of complex numbers which eliminates a degeneracy of these numbers. In the opposite direction, *the field of complex numbers is the center of the algebra of quantions.*

Of the two restrictions from quantions to complex numbers, the algebraic one, relation (5.7), is evidently the appropriate one for our purposes. The reason is that quantionic wave functions are meant to generalize complex wave functions. In support of this intuitive conclusion, let us consider an 'infinitesimal extension' of the field of complex numbers as *the set $\tilde{\mathbb{C}}$ of quantions which differ from complex numbers by at most an infinitesimal object:*

$$Q = zI + \varepsilon Q_0 \in \tilde{\mathbb{C}} \subset \mathcal{L}. \tag{5.8}$$

The parameter ε is infinitesimal, and the quantion

$$Q_0 = \begin{pmatrix} u & v & 0 & 0 \\ v & -u & 0 & 0 \\ 0 & 0 & u & v \\ 0 & 0 & v & -u \end{pmatrix} \tag{5.9}$$

is traceless. It can be normalized without loss of generality. A normalization condition which suggests itself is

$$Q_0^\dagger Q_0 = I. \tag{5.10}$$

The rationale for normalization is that a general traceless quantion is defined by six real parameters, so that the adjunction of a seventh parameter, ε, must be compensated by some real condition.

In the lambda expansion (2.18), a quantion of type (5.8) is of the form

$$Q = z\Lambda_0 + \varepsilon\,(a\Lambda_1 + b\Lambda_2 + c\Lambda_3) \in \tilde{\mathbb{C}}. \tag{5.11}$$

The normalization condition (5.10) yields

$$a^* a + b^* b + c^* c = 1. \tag{5.12}$$

A digression into physics

The physical applications of quantions are formally developed in Part II, but a brief digression into physics will bring out the importance of the neighborhood $\tilde{\mathbb{C}}$ of the field $\mathbb{C}$ of complex numbers. To this end, consider the following inclusions:

$$\mathbb{C} \subset \tilde{\mathbb{C}} \subset \mathcal{L}.$$

An intermediate step in the quantionic generalization of quantum mechanics is based on the following points:

Observations concerning quantum mechanics:
(a) Quantum mechanical amplitudes are represented by complex numbers, $\psi \in \mathbb{C}$.
(b) Schrödinger's equation for the amplitude ψ is postulated.
(c) Born's interpretation assigns the physical meaning of probability density to the norm $\psi^* \psi$.

Observations concerning the quantionic approach:
(a) We generalize amplitudes from complex numbers to arbitrary quantions:

$$\mathbb{C} \ni \psi \longrightarrow Q \in \mathcal{L}.$$

(b) Zovko's interpretation of the norm $Q^\dagger Q$ of a quantionic amplitude generalizes Born's interpretation of the norm $\psi^*\psi$ of a complex amplitude.

(c) No equation for the amplitude Q is postulated, but such an equation is derived as a theorem from Zovko's interpretation.

The intermediate step:

(a) We initially generalize complex amplitudes to quantionic amplitudes that are infinitesimally close to complex numbers:

$$\mathbb{C} \ni \psi \longrightarrow Q = \psi I + \varepsilon Q_0 \in \tilde{\mathbb{C}}.$$

(b) Zovko's interpretation of $Q^\dagger Q$ yields the Schrödinger equation for ψ as an exact theorem (derived in Chapter 15).

The algebraic norm $Q^\dagger Q$ of a quantion Q which differs infinitesimally from a complex number ψ is

$$
\begin{aligned}
Q^\dagger Q &= (\psi I + \varepsilon Q_0)^\dagger (\psi I + \varepsilon Q_0) \\
&= \left(\psi^* I + \varepsilon Q_0^\dagger\right)(\psi I + \varepsilon Q_0) \\
&= \psi^*\psi + \varepsilon \left(\psi Q_0^\dagger + \psi^* Q_0\right).
\end{aligned}
\tag{5.13}
$$

The matrix $\left(\psi Q_0^\dagger + \psi^* Q_0\right)$ by which $Q^\dagger Q$ differs from $\psi^*\psi$ is traceless and Hermitian. We thus note that both tracelessness and Hermiticity are physically important in the transition to quantionic physics.

The metric norm

We observe that the matrix (5.4) has been introduced as a geometric transformation. These transformation form a group, so that the inverse Q^{-1} is an algebraically essential object conceptually independent of Hermitian conjugation and of the algebraic norm.

Let us first think of Q as an arbitrary matrix (not necessarily a quantion). Its inverse, Q^{-1}, is the matrix of the minors of Q divided by the determinant of Q. Hence, after cancellation of common polynomial factors, if there are any, the inverse is of the form

$$Q^{-1} = \frac{1}{m(Q)} Q^\#, \tag{5.14}$$

where $m(Q)$ is a polynomial and $Q^{\#}$ a matrix of polynomials. We shall refer to $Q^{\#}$ as the "dual" of Q — or "metric dual", to avoid confusion with the "algebraic dual" introduced earlier.

For the matrix Q given by (5.4), we have

$$
\begin{aligned}
Q^{-1} &= \begin{pmatrix} x & -y & 0 & 0 \\ y & x & 0 & 0 \\ 0 & 0 & x & -y \\ 0 & 0 & y & x \end{pmatrix}^{-1} \\
&= \frac{1}{x^2 + y^2} \begin{pmatrix} x & y & 0 & 0 \\ -y & x & 0 & 0 \\ 0 & 0 & x & y \\ 0 & 0 & -y & x \end{pmatrix},
\end{aligned}
\tag{5.15}
$$

which yields

$$
m(Q) = x^2 + y^2.
\tag{5.16}
$$

Since this is a homogeneous second order polynomial, it may be interpreted as a metric norm.

Comparison with the previous results yields

$$
\left.\begin{aligned}
Q^{\#} &= Q^{\dagger}, \\
m(Q) &= A(Q).
\end{aligned}\right\}
\tag{5.17}
$$

These identities are theorems specific to the complex numbers, and, as shown in the next section, to the quaternions. They are not universally true, however. This is why we say that the complex numbers are degenerate.

An observation: The algebraic limit (5.7) and the geometric limit (5.4) are both representations of the complex numbers, but they are very different subalgebras of $\mathcal{L}$: The algebraic limit is the center of $\mathcal{L}$, the geometric limit is not.

5.2 Comparison with quaternions

Quaternions may be viewed as three 'intertwined' complex numbers that share the same diagonal elements:

$$Q = \begin{pmatrix} w & -x & -y & -z \\ x & w & z & -y \\ y & -z & w & x \\ z & y & -x & w \end{pmatrix}. \tag{5.18}$$

To see why, take any of the six 2×2 sub-matrices on the diagonal: Each is of the form (5.3).

The Hermitian conjugate is

$$Q^{\dagger} = \begin{pmatrix} w & x & y & z \\ -x & w & -z & y \\ -y & z & w & -x \\ -z & -y & x & w \end{pmatrix},$$

The inverse yields the metric dual

$$Q^{-1} = \frac{1}{M(Q)} Q^{\#} = \begin{pmatrix} w & -x & -y & -z \\ x & w & z & -y \\ y & -z & w & x \\ z & y & -x & w \end{pmatrix}^{-1}$$

$$= \frac{1}{w^2 + x^2 + y^2 + z^2} \begin{pmatrix} w & x & y & z \\ -x & w & -z & y \\ -y & z & w & -x \\ -z & -y & x & w \end{pmatrix}.$$

Hence, the algebraic norm and the metric norm are

$$\begin{aligned} M(Q) &= Q^{\#}Q = (w^2 + x^2 + y^2 + z^2)\,I, & (5.19) \\ A(Q) &= Q^{\dagger}Q = (w^2 + x^2 + y^2 + z^2)\,I. & (5.20) \end{aligned}$$

Thus, the relations (5.17) also hold for quaternions, which means that the quaternions are as degenerate as the complex numbers.

Another representation of quaternions

Consider now the special quantions of the form

$$
Q = \begin{pmatrix} a & -b^* & 0 & 0 \\ b & a^* & 0 & 0 \\ 0 & 0 & a & -b^* \\ 0 & 0 & b & a^* \end{pmatrix},
\tag{5.21}
$$

where a and b are arbitrary complex numbers.

The totality of these matrices is algebraically stable. For addition, stability is obvious: the sum of two matrices of type (5.21) is of the same type. To verify the stability of multiplication, consider the product of two such matrices in reduced form:

$$
\begin{pmatrix} a & -b^* \\ b & a^* \end{pmatrix} \begin{pmatrix} u & -v^* \\ v & u^* \end{pmatrix} = \begin{pmatrix} (au - vb^*) & -(va^* + bu)^* \\ (va^* + bu) & (au - vb^*)^* \end{pmatrix}.
$$

The two factors and the product are all of the same type, which is the definition of stability.

The inverse of a non-vanishing matrix of type (5.21) also exists, and is of the same type:

$$
\begin{pmatrix} a & -b^* \\ b & a^* \end{pmatrix}^{-1} = \frac{1}{aa^* + bb^*} \begin{pmatrix} (a^*) & -(-b)^* \\ (-b) & (a^*)^* \end{pmatrix}.
$$

Associativity is a consequence of the associativity of the matrix product.

The product is non-commutative:

$$
\begin{pmatrix} a & -b^* \\ b & a^* \end{pmatrix} \begin{pmatrix} u & -v^* \\ v & u^* \end{pmatrix} - \begin{pmatrix} u & -v^* \\ v & u^* \end{pmatrix} \begin{pmatrix} a & -b^* \\ b & a^* \end{pmatrix}
$$

$$
= \begin{pmatrix} bv^* - vb^* & ub^* - av^* + a^*v^* - b^*u^* \\ va^* - bu^* - av + bu & vb^* - bv^* \end{pmatrix}
$$

Hence, the matrices of type (5.21) form a non-commutative associative division algebra in four real dimensions. There is only one such algebra. It is the field of quaternions.

The representations (5.18) and (5.21) are very different: The first is not an element of $\mathcal{L}$, the second is.

Chapter 6

The Norms of Quantions

In the standard representations of complex numbers, quaternions and octonions as vectors in linear spaces of two, four, and eight dimensions,

$$\left.\begin{array}{l} z = x + iy, \\ q = u + ix + jy + kz, \\ f = u + e_1 x^1 + \cdots + e_7 x^7 \end{array}\right\} \qquad (6.1)$$

the respective algebraic and metric norms are

$$\left.\begin{array}{l} A(z) = M(z) = z^* z = x^2 + y^2, \\ A(q) = M(q) = q^* q = u^2 + x^2 + y^2 + z^2, \\ A(f) = M(f) = u^2 + \left(x^1\right)^2 + \cdots + \left(x^7\right)^2. \end{array}\right\} \qquad (6.2)$$

These positive definite Pythagorean expressions are characteristic of division algebras. They contain no indication of how they could be generalized to the algebra of quantions, where $A(q) \neq M(q)$.

Since quantions are matrices, it is in the matrix representations of complex numbers that analogies are to be found. Thus, Chapter 5 suggests the expressions

$$A(Q) = Q^\dagger Q, \qquad (6.3)$$

$$M(Q) = Q^\# Q. \qquad (6.4)$$

for the two norm functions of quantions. These expressions are also valid for complex numbers and quaternions, but not for octonions (which have no matrix representation for not being associative).

6.1   The metric norm

As in the case of complex numbers and quaternions in their matrix representations, we define the dual $Q^{\#}$ (read "Q sharp") and the scalar metric norm $m(Q)$ of a quantion Q by way of the formal inverse:

$$Q^{-1} = \frac{1}{m(Q)}Q^{\#}. \tag{6.5}$$

The "formal inverse" is the matrix Q^{-1} of functions of the complex variables q_1 to q_4, even though the quantion Q^{-1} itself does not exist if $m(Q) = 0$.

Since the right-hand side of (6.5) is a ratio, the numerator and the denominator can be rescaled by an arbitrary common factor. To fix them, it suffices to require that $I^{\#} = I$, and that the polynomial $m(Q)$ be mutually prime to every matrix elements of $Q^{\#}$. Moreover, since $m(Q)$ is to be interpreted as a metric, it must be a homogeneous second order polynomial. It was shown in [10] that the only algebra in which the last condition is satisfied without degeneracy is the algebra of quantions.

Simpler but equivalent definitions of $Q^{\#}$ and $m(Q)$ may be stated in the block-diagonal form of quantions, where the inverse is

$$Q^{-1} = \begin{pmatrix} A & 0 \\ 0 & A \end{pmatrix}^{-1} = \begin{pmatrix} A^{-1} & 0 \\ 0 & A^{-1} \end{pmatrix} = \frac{1}{\det A} \begin{pmatrix} A^{\#} & 0 \\ 0 & A^{\#} \end{pmatrix}.$$

In this expression, the elements of $A^{\#}$ and $\det A$ are automatically mutually prime because $A = Q_{red}$ is an arbitrary 2×2 matrix.

Definition 20 *The **metric norm** of a quantion $Q \in \mathcal{L}$ is defined, as needed, as a scalar function denoted by $m(*)$,*

$$m(Q) \stackrel{def}{=} \det Q_{red}, \tag{6.6}$$

or as a 4×4 matrix function denoted by $M(*)$,

$$M(Q) \stackrel{def}{=} m(Q)I. \tag{6.7}$$

Since Q_{red} is a 2×2 matrix, $m(Q)$ is automatically a homogeneous polynomial of the second order.

The metric norm being thus defined, the general definition of the metric dual follows from relations (6.5) and (6.6):

Definition 21 *The **(metric) dual** $Q^{\#} \in \mathcal{L}$ of a quantion $Q \in \mathcal{L}$ is the quantion*

$$Q^{\#} = m(Q) Q^{-1} \equiv M(Q) Q^{-1}. \tag{6.8}$$

Thus, clearly,

$$M(Q) = QQ^{\#} = Q^{\#}Q. \tag{6.9}$$

The justification for calling $Q^{\#}$ the dual of Q is the formal analogy between the norm (6.9) of a quantion and the norm

$$g_{\mu\nu} v^{\mu} v^{\nu} = v^{\mu} v_{\mu}, \tag{6.10}$$

of a vector v^{μ} in a linear metric space.

This analogy extends to the scalar product: An arbitrary fixed vector p^{μ} defines a linear functional in the linear space of vectors q^{μ}. It is the mapping $p_{\mu} q^{\mu}$ of q^{μ} into the field of real numbers.

Similarly, an arbitrary fixed quantion P defines a linear functional in the algebra of quantions Q. It is the mapping $\frac{1}{2}(P^{\#}Q + Q^{\#}P)$ of Q into the field of complex numbers. The symmetrization is necessary because the product of matrices does not commute.

In the full matrix form, the metric dual of a quantion is

$$\begin{pmatrix} q_1 & q_3 & 0 & 0 \\ q_2 & q_4 & 0 & 0 \\ 0 & 0 & q_1 & q_3 \\ 0 & 0 & q_2 & q_4 \end{pmatrix}^{\#} = \begin{pmatrix} q_4 & -q_3 & 0 & 0 \\ -q_2 & q_1 & 0 & 0 \\ 0 & 0 & q_4 & -q_3 \\ 0 & 0 & -q_2 & q_1 \end{pmatrix}. \tag{6.11}$$

The sharp operation yields an essentially new quantion. It is new in the sense that $Q^{\#}$ is not related to Q by a similarity transformation. This is easily verified in $\mathcal{L}_{red}$, where the equation

$$AS = SA,$$

viewed as an identity in A, has only the trivial solution $S = sI$.

6.2   The algebraic norm

The algebraic norm (6.3) is meaningful in any matrix algebra. Hence, in particular, it exists in the algebra of quantions:

Definition 22 *The **algebraic norm** $A(Q)$ of a quantion Q is defined as*

$$A(Q) = Q^\dagger Q. \tag{6.12}$$

*The function $A(*)$ is referred to as the **algebraic norm function**.*

In general, Q and $Q^\dagger$ do not commute, so that $A(Q^\dagger) \neq A(Q)$.

The product $Q^\dagger Q$ has been selected instead of the opposite product $QQ^\dagger$ for agreement with the formalism which is standard in Hilbert space, where operators act on vectors from the left. We shall nevertheless introduce a name for the opposite product:

Definition 23 *The **alternative algebraic norm** of a quantion Q, denoted by $B(Q)$, is defined as*

$$B(Q) = QQ^\dagger = A\left(Q^\dagger\right). \tag{6.13}$$

The following theorem plays an essential role in the physical interpretation of quantions.

Theorem 24 *For every quantion $Q \in \mathcal{L}$, the algebraic norm $A(Q)$ belongs to $\mathcal{L}^+$.*

Proof. This means that: (a) $A(Q)$ is a Hermitian quantion, (b) its trace is positive definite, and (c) its determinant is real and non-negative.

Proof of part (a):

$$[A(Q)]^\dagger = \left[Q^\dagger Q\right]^\dagger = Q^\dagger Q^{\dagger\dagger} = Q^\dagger Q = A(Q).$$

Proof of part (b):

$$Tr(A(Q)) = 2(q_1^* q_1 + q_2^* q_2 + q_3^* q_3 + q_4^* q_4) \geqslant 0. \tag{6.14}$$

This expression is positive definite because it vanishes only if all four variables vanish simultaneously.

Proof of part (c):

$$A\left(Q\right) = Q^{\dagger}Q = \begin{pmatrix} A & 0 \\ 0 & A \end{pmatrix}^{\dagger} \begin{pmatrix} A & 0 \\ 0 & A \end{pmatrix}$$

$$= \begin{pmatrix} A^{\dagger} & 0 \\ 0 & A^{\dagger} \end{pmatrix} \begin{pmatrix} A & 0 \\ 0 & A \end{pmatrix} = \begin{pmatrix} A^{\dagger}A & 0 \\ 0 & A^{\dagger}A \end{pmatrix}.$$

Hence,

$$\det\left(A\left(Q\right)\right) = \left[\det\left(A^{\dagger}A\right)\right]^{2},$$

and

$$\det\left(A^{\dagger}A\right) = \det\left(A^{\dagger}\right)\det\left(A\right) = \left[\det\left(A\right)\right]^{*}\det\left(A\right) \geqslant 0. \qquad (6.15)$$

This proves that $\det\left(A\left(Q\right)\right)$ is real and non-negative, but since the determinant

$$\det\left(A\right) = \det \begin{pmatrix} q_{1} & q_{3} \\ q_{2} & q_{4} \end{pmatrix} = q_{1}q_{4} - q_{2}q_{3} \qquad (6.16)$$

is a complex number which may vanish even if the variables themselves do not vanish, it follows that $\det\left(A\left(Q\right)\right)$ may vanish for some $Q \neq 0$. In other words, it is not positive definite. ∎

6.3 The fundamental theorem

Since the algebraic norm of an arbitrary quantion is a Hermitian matrix while its metric norm is essentially a complex number, the two cannot be equal:

$$A\left(Q\right) \neq M\left(Q\right).$$

Thus, unlike complex numbers and quaternions, quantions are not degenerate.

As an aside, let us verify that the quantions of type (5.21) — whose algebra has been shown to be isomorphic to the field of quaternions — are indeed degenerate. In terms of the sub-matrices

$$A = \begin{pmatrix} a & -b^{*} \\ b & a^{*} \end{pmatrix},$$

we have

$$A^\dagger = \begin{pmatrix} a^* & b^* \\ -b & a \end{pmatrix},$$
$$A^\dagger A = (aa^* + bb^*) I,$$

and

$$A^{-1} = \frac{1}{aa^* + bb^*} \begin{pmatrix} a^* & b^* \\ -b & a \end{pmatrix},$$
$$A^\# = \begin{pmatrix} a^* & b^* \\ -b & a \end{pmatrix}.$$

Hence $Q^\dagger = Q^\#$ and $A(Q) = M(Q)$, proving the degeneracy.

The two norms of a quantion Q are related by the following theorem, which is considered fundamental due to the physical meanings of the two norms (see page 18).

Theorem 25 *The metric norm of the algebraic norm and the algebraic norm of the metric norm are identical:*

$$MA(Q) \equiv AM(Q). \tag{6.17}$$

Proof. In the reduced representation, we have the identity

$$\det\left(Q_{red}^\dagger Q_{red}\right) \equiv [\det(Q_{red})]^* \det(Q_{red}).$$

The left hand side is the metric norm of the algebraic norm:

$$\det\left(Q_{red}^\dagger Q_{red}\right) = MA(Q).$$

The right hand side is the algebraic norm of the metric norm:

$$[\det(Q_{red})]^* \det(Q_{red}) = AM(Q).$$

This proves the identity (6.17). ∎

This result suggests a new concept:

Definition 26 *The **double norm** of a quantion is the non-negative real number $n(Q)$ defined as*

$$n(Q) I \overset{def}{=} MA(Q) \equiv AM(Q) \in \mathbb{R}^+ I. \tag{6.18}$$

We observe that the double norm is a fourth-order polynomial.

Chapter 7

The Commutation Theorems

Five operations have been defined on quantions:
— the *algebraic product,*
— the *sharp operation,*
— the *dagger operation,*
— the *algebraic norm,*
— the *metric norm.*

As each of these operations yields a quantion, any two of them may be performed in sequence (as in Theorem 25).

The "commutation theorems" derived in this chapter are about the properties of two consecutive operations and about the commutator and anti-commutator of the algebraic product.

Commutativity of different operations

To present these theorems systematically, all pairs of operations are tabulated in the 5×5 array (7.1). The product of matrices is represented in the table by $\times$, that is $P \times Q \equiv PQ$.

The fifteen entries in parentheses are pointers to the numbered discussions which follow. The rule of precedence is that the operations in the rows are performed before the operations in the columns. Thus, for example, the expression $(PQ)^{\#}$ must be viewed as $(P \times Q)^{\#}$, the first operation to be performed being $\times$. The table entry for $(\times, \#)$

is (5), and the discussion (5) on page (89) returns the commutation identity followed by a proof. In the opposite order, $(\#, \times)$, the table entry is also (5), and the identity is the same, but read from right to left, $P^{\#}Q^{\#} = (QP)^{\#}$.

	$\times$	$\dagger$	$\#$	M	A
$\times$	(1)	(4)	(5)	(7)	(8)
$\dagger$	(4)	(2)	(6)	(9)	(10)
$\#$	(5)	(6)	(3)	(11)	(12)
M	(7)	(9)	(11)	(13)	(15)
A	(8)	(10)	(12)	(15)	(14)

(7.1)

We shall now discuss all fifteen entries sequentially.

(1) Two products

The algebra of quantions is associative. This means that if three arbitrary quantions are given in some sequence, their two possible successive products are identical — which implies that the parentheses may be dropped without ambiguity:

$$QRS \overset{def}{=} (QR)\,S \equiv Q\,(RS)\,. \tag{7.2}$$

Associativity being a native property of the algebra of matrices, it need not be proved separately for quantions.

(2) Two daggers

The dagger operation is its own inverse. In other words, Hermitian conjugation is involutive:

$$Q^{\dagger\dagger} = Q. \tag{7.3}$$

The reason is that

$$Q^{\dagger} \overset{def}{=} (Q^*)^{\top} = \left(Q^{\top}\right)^*, \tag{7.4}$$

and both the transpose of a matrix and the complex conjugation of its elements commute and are involutive, $Q^{**} = Q$, $Q^{\top\top} = Q$.

(3) Two sharps

The sharp operation is its own inverse. In other words, the dual is involutive:

$$Q^{\#\#} = Q. \tag{7.5}$$

To verify this identity, we apply the sharp operator to both sides of the defining relation (6.8). Taking into account that the determinant of the product of two matrices is the product of the two determinants, one obtains the following sequence of relations

$$
\begin{aligned}
Q^{\#\#} &= \left[(\det Q_{red}) Q^{-1}\right]^{\#} = (\det Q_{red}) \left(Q^{-1}\right)^{\#} \\
&= (\det Q_{red}) \left(\det Q_{red}^{-1}\right) \left(Q^{-1}\right)^{-1} = Q.
\end{aligned}
$$

This proves the assertion.

(4) Product and dagger

The algebraic product of quantions is preserved by the dagger operation, but in the opposite order:

$$(QR)^{\dagger} = R^{\dagger} Q^{\dagger}. \tag{7.6}$$

This type of identity is referred to as an anti-automorphism.

The reason Hermitian conjugation is an anti-automorphism is that the transpose in the definition (7.4) is an anti-automorphism, while complex conjugation is an automorphism: $(QR)^{*} = Q^{*} R^{*}$. It is evident that an even number of successive anti-automorphisms is an automorphism, and that an odd number of successive anti-automorphisms is an anti-automorphism.

(5) Product and sharp

The dual is an anti-automorphism:

$$(QP)^{\#} = P^{\#} Q^{\#}. \tag{7.7}$$

To verify this identity, we apply the definition (6.8) to a product of quantions:

$$(QP)^{\#} = (\det (QP)_{red}) (QP)^{-1}.$$

Since the inverse is an anti-automorphism, we have

$$(QP)^{\#} = (\det Q_{red})\,(\det P_{red})\, P^{-1}Q^{-1} = P^{\#}Q^{\#}.$$

(6) Dagger and sharp

Hermitian conjugation commutes with the dual:

$$Q^{\#\dagger} = Q^{\dagger\#}. \tag{7.8}$$

It is understood that the two operations are performed in the indicated order. Thus, for example, $Q^{\#\dagger} \overset{def}{=} (Q^{\#})^{\dagger}$.

To verify the identity (7.8), it suffices to apply the definition (6.8) of the dual to the Hermitian conjugate of a quantion:

$$\left(Q^{\dagger}\right)^{\#} = \left(\det Q^{\dagger}_{red}\right)\left(Q^{\dagger}\right)^{-1}.$$

Since

$$\left(Q^{\dagger}\right)^{-1} = \left(Q^{-1}\right)^{\dagger},$$

one obtains

$$
\begin{aligned}
\left(Q^{\dagger}\right)^{\#} &= (\det Q_{red})^{*}\,(Q^{-1})^{\dagger} \\
&= [(\det Q_{red})\,(Q^{-1})]^{\dagger} = \left(Q^{\#}\right)^{\dagger}.
\end{aligned}
$$

(7) Product and metric norm

This case is important enough to be the subject of a theorem:

Theorem 27 *The metric norm of the product of (any number of) quantions is the product of the metric norms of the individual factors.*

$$M\,(PQ) = M\,(P)\,M\,(Q). \tag{7.9}$$

Proof. By relations (6.9) and (7.7):

$$M\,(PQ) = (PQ)^{\#}\,(PQ) = Q^{\#}P^{\#}PQ = Q^{\#}M\,(P)\,Q.$$

Since the metric norm is a complex number, $M(P)$ commutes with all quantions. Hence,

$$M(PQ) = M(P)Q^{\#}Q = M(P)M(Q),$$

proving relation (7.9). ∎
  This result implies

$$M(PQ) = M(QP).$$

(8) Product and algebraic norm

Of the fifteen pairs of operations, this is the only one which seems to have no interesting property. The reason is that the expression

$$A(PQ) = (PQ)^{\dagger}(PQ) = Q^{\dagger}P^{\dagger}PQ = Q^{\dagger}A(P)Q$$

is not simply related to either of the expressions

$$\begin{aligned} A(P)A(Q) &= P^{\dagger}PQ^{\dagger}Q, \\ A(Q)A(P) &= Q^{\dagger}QP^{\dagger}P, \end{aligned}$$

or to their linear combinations.
  The following observation from the history of mathematics is meant to put this 'unpleasant' result in perspective:
  For any two pairs of real numbers, $\{u, v\}$ and $\{x, y\}$, the identity

$$\left(u^2 + v^2\right)\left(x^2 + y^2\right) \equiv (ux - vy)^2 + (uy + vx)^2, \tag{7.10}$$

which was already known to Diophantos in the third century AD, may be viewed as a generalization of the trivial identity

$$u^2 x^2 \equiv (ux)^2. \tag{7.11}$$

The question thus arose whether it could be generalized to more variables, that is, to identities of the type

$$\left(u_1^2 + u_2^2 + \cdots + u_n^2\right)\left(x_1^2 + x_2^2 + \cdots + x_n^2\right) \equiv \text{Sum of } n \text{ squares.}$$

It was discovered in the 19th century that such identities exist only for $n = 1, 2, 4$ and 8. The subsequent interpretation of this result was that they correspond to the more conceptual identity

$$A(uv) = A(u) A(v),   \tag{7.12}$$

which is satisfied only by the algebraic norms in the the four division algebras. Due to the degeneracy of these algebras, that is, $A(*) = M(*)$, the identity (7.12) may also be interpreted as

$$M(uv) = M(u) M(v).   \tag{7.13}$$

For quantions, which are not degenerate, the relation (7.12) does not hold, while relation (7.13) does.

(9) Metric norm and dagger

The metric norm of the Hermitian conjugate is the Hermitian conjugate of the metric norm:

$$M\left(Q^\dagger\right) = [M(Q)]^\dagger.   \tag{7.14}$$

Proof:

$$M\left(Q^\dagger\right) = Q^{\dagger\#}Q^\dagger = Q^{\#\dagger}Q^\dagger = \left(QQ^\#\right)^\dagger = [M(Q)]^\dagger = [m(Q)]^* I.$$

(10) Algebraic norm and dagger

The algebraic norm of a quantion is invariant under Hermitian conjugation. In other words, $A(Q)$ is Hermitian:

$$[A(Q)]^\dagger = \left(Q^\dagger Q\right)^\dagger = Q^\dagger Q = A(Q).$$

The same is true of the alternative algebraic norm

$$[B(Q)]^\dagger = \left[A\left(Q^\dagger\right)\right]^\dagger = \left(QQ^\dagger\right)^\dagger = Q^{\dagger\dagger}Q^\dagger = QQ^\dagger = B(Q).$$

(11) Metric norm and sharp

The metric norm of a quantion and the metric norm of its dual are identical:

$$M\left(Q^{\#}\right) = Q^{\#\#}Q^{\#} = QQ^{\#} = M\left(Q\right). \qquad (7.15)$$

(12) Algebraic norm and sharp

The algebraic norm of the dual of a quantion is the dual of the alternative algebraic norm:

$$A\left(Q^{\#}\right) = [B\left(Q\right)]^{\#}. \qquad (7.16)$$

Proof:

$$A\left(Q^{\#}\right) = Q^{\#\dagger}Q^{\#} = \left(QQ^{\dagger}\right)^{\#} = [B\left(Q\right)]^{\#}.$$

(13) Two metric norms

Since the metric norm of a quantion is a quantion, the norm of a norm, $M\left(M\left(Q\right)\right)$, written for short as $MM\left(Q\right)$, is a meaningful and well-defined concept.

The metric norm of a metric norm is the square of the metric norm:

$$MM\left(Q\right) = [M\left(Q\right)]^{2}. \qquad (7.17)$$

Since $M\left(Q\right)$ is a complex number, relation (7.17) is a special case of the following more general identity: The metric norm of a central quantion Q is Q^{2}.

Since a central quantion is of the form

$$Q = zI = \begin{pmatrix} z & 0 & 0 & 0 \\ 0 & z & 0 & 0 \\ 0 & 0 & z & 0 \\ 0 & 0 & 0 & z \end{pmatrix},$$

we have

$$\det Q_{red} = z^{2}.$$

Hence:

$$M\left(Q\right) = (\det Q_{red})\, I = z^{2}I = Q^{2}.$$

(14) Two algebraic norms

The algebraic norm of an algebraic norm is the square of the algebraic norm:

$$AA(Q) = [A(Q)]^2. \tag{7.18}$$

This relation is a special case of the following more general identity: The algebraic norm of a Hermitian quantion H is H^2.

Since $H^\dagger = H$, we have

$$A(H) = H^\dagger H = HH = H^2. \tag{7.19}$$

(15) Algebraic norm and metric norm

By Theorem 25, the metric and algebraic norms commute, which we write as

$$AM(Q) \equiv MA(Q).$$

Since this is a fundamental identity in the algebra of quantions, let us prove it again by expanding the definition of $AM(Q)$:

$$
\begin{aligned}
AM(Q) &= A(I \det Q_{red}) = (I \det Q_{red})^\dagger (I \det Q_{red}) \\
&= \left(\det Q_{red}^\dagger\right)(\det Q_{red}) I = \det\left(Q_{red}^\dagger Q_{red}\right) I \\
&= MA(Q).
\end{aligned}
$$

As an example of the commutativity of the two norms, let us prove that the double norm $n(Q)$ is a non-negative real number by computing it in both orderings.

1. The definition $n = AM$:

$$n(Q) = A(m(Q)) = (m(Q))^* (m(Q)) \in \mathbb{R}^+.$$

2. The definition $n = MA$:

$$n(Q) = m(A(Q)) = m\left(Q^\dagger Q\right) = (m(Q))^* (m(Q)).$$

These two results are identical.

The commutator and the anticommutator

Lemma 28 *Two Hermitian quantions, $R, S \in \mathcal{L}^h$, commute*

$$[R, S] = 0$$

if and only if

$$S = \lambda R + \mu I$$

for arbitrary real coefficients λ, μ.

Proof. Let us write

$$\begin{aligned} R &= \vec{r} \cdot \vec{\sigma} + rI, \\ S &= \vec{s} \cdot \vec{\sigma} + sI, \end{aligned}$$

where R and S stand for R_{red} and S_{red}. Then,

$$[R, S] = 2i \left(\vec{r} \times \vec{s} \right) \cdot \vec{\sigma},$$

which vanishes if and only if

$$\vec{s} = \lambda \vec{r}.$$

while s is arbitrary, proving the assertion. ∎

Lemma 29 *For Hermitian quantions, $R, S \in \mathcal{L}^h$, two types of solutions satisfy the anticommutator equation*

$$[R, S]_+ = 0.$$

Case (1):

$$R = \begin{pmatrix} a & c - id \\ c + id & -a \end{pmatrix} \quad and \quad S = \begin{pmatrix} \alpha & \gamma - i\delta \\ \gamma + i\delta & -\alpha \end{pmatrix} \qquad (7.20)$$

under the Euclidian orthogonality condition

$$a\alpha + c\gamma + d\delta = 0. \qquad (7.21)$$

Case (2):

$$R = \begin{pmatrix} a + b & c - id \\ c + id & a - b \end{pmatrix} \quad and \quad S = \lambda R^{\#} \qquad (7.22)$$

for arbitrary real λ, under the Minkowskian null condition

$$a^2 - b^2 - c^2 - d^2 = 0. \qquad (7.23)$$

Proof. Note: R and S stand for R_{red} and S_{red}. The proof consists of two parts: We first prove that there are two types of solutions, and then we characterize the matrices R and S in each case.

Part one of the proof:

Let R and S be two arbitrary Hermitian matrices

$$\left. \begin{aligned} R &= \begin{pmatrix} a & c-id \\ c+id & b \end{pmatrix}, \\ S &= \begin{pmatrix} \alpha & \gamma-i\delta \\ \gamma+i\delta & \beta \end{pmatrix}. \end{aligned} \right\} \tag{7.24}$$

The vanishing of their anticommutator,

$$\begin{aligned} 0 &= [R,S]_+ \\ &= \begin{pmatrix} 2(a\alpha+c\gamma+d\delta) \\ (a+b)(\gamma+i\delta)+(c+id)(\alpha+\beta) \end{pmatrix. \\ &\qquad \left| \begin{matrix} (a+b)(\gamma+i\delta)+(c+id)(\alpha+\beta) \\ 2(a\alpha+c\gamma+d\delta) \end{matrix} \right) \end{aligned} \tag{7.25}$$

is equivalent to the following system of homogeneous equations

$$\begin{aligned} a\alpha+c\gamma+d\delta &= 0 \tag{7.26} \\ b\beta+c\gamma+d\delta &= 0 \tag{7.27} \\ (a+b)\gamma+c(\alpha+\beta) &= 0 \tag{7.28} \\ (a+b)\delta+d(\alpha+\beta) &= 0 \tag{7.29} \end{aligned}$$

Solutions for $\alpha, \beta, \gamma, \delta$ in terms of a, b, c, d (and vice-versa) exist only if the determinant vanishes:

$$D = \det \begin{pmatrix} a & 0 & c & d \\ 0 & b & c & d \\ c & c & a+b & 0 \\ d & d & 0 & a+b \end{pmatrix} = (a+b)^2\left(ab-c^2-d^2\right) = 0.$$

Thus, in terms of the invariant functions Tr and $\det$, the equation $[R,S]_+ = 0$ has at least one solutions if and only if

$$(TrR)^2 \det R = (TrS)^2 \det S = 0. \tag{7.30}$$

Since (by the Cayley-Hamilton theorem applied to 2×2 matrices) the case of $TrR = \det R = 0$ implies $R = 0$ (and similarly for S), the condition (7.30) can be satisfied in exactly four different ways:

$$
\begin{aligned}
TrR &= TrS = 0, & (7.31) \\
\det R &= \det S = 0, & (7.32) \\
TrR &= \det S = 0, & (7.33) \\
\det R &= TrS = 0. & (7.34)
\end{aligned}
$$

Relations (7.31) and (7.32) correspond to case(1) and case (2) respectively. In components, relations (7.33) read

$$
\begin{aligned}
a + b &= 0, \\
\alpha\beta - \gamma^2 - \delta^2 &= 0.
\end{aligned}
$$

With relations (7.28) and (7.29), they yield

$$
\begin{aligned}
c\,(\alpha + \beta) &= -\gamma\,(a + b), \\
d\,(\alpha + \beta) &= -\delta\,(a + b).
\end{aligned}
$$

Hence:

$$
\begin{aligned}
c &= d = 0, \\
b &= -a.
\end{aligned}
$$

Substitution of these values into equations (7.26) and (7.27) yields

$$
a\alpha = a\beta = 0
$$

It follows that $a = 0$, which implies the trivial solution $R = 0$, or $\alpha = \beta = \gamma = \delta = 0$, that is $S = 0$. The same conclusion holds for the case (7.34). Thus, the cases (1) and (2) are the only non-trivial solutions.

Part two of the proof:

In case (1), the traces of the matrices (7.24) vanish,

$$
a + b = \alpha + \beta = 0,
$$

but the determinants do not. This yields the expressions (7.20) for the matrices R and S, while the commutator condition (7.25) simplifies to

$$(a\alpha + c\gamma + d\delta)\, I = 0,$$

which is equivalent to the orthogonality condition (7.21).

In case (2), the determinants of the matrices (7.24) vanish

$$ab - c^2 - d^2 = \alpha\beta - \gamma^2 - \delta^2 = 0,$$

but the traces do not. Taking the difference of equations (7.26) and (7.27) yields

$$a\alpha - b\beta = 0.$$

Hence:

$$
\begin{aligned}
\alpha &= \lambda b, \\
\beta &= \lambda a,
\end{aligned}
$$

for an arbitrary real factor λ. Substitution of these relations into (7.27) and (7.28) yields

$$
\begin{aligned}
(a+b)\,\gamma + \lambda c\,(a+b) &= 0, \\
(a+b)\,\delta + \lambda d\,(a+b) &= 0.
\end{aligned}
$$

Since $a + b \neq 0$, we have

$$
\begin{aligned}
\gamma &= -\lambda c, \\
\delta &= -\lambda d.
\end{aligned}
$$

Substitution of the solutions for α to δ into the matrix S in (7.24) yields

$$S = \lambda \begin{pmatrix} b & -(c - id) \\ -(c + id) & a \end{pmatrix} \equiv \lambda R^{\#}.$$

The vanishing of the determinant assumes the geometrically meaningful form (7.23) of a null vector in Minkowski space if we rename the variables: $a + b$ instead of a, and $a - b$ instead of b. The matrix R then assumes the form (7.22), while the traces are

$$Tr R = Tr S = 2a.$$

Hence, the non-vanishing of the trace implies that the vectors r^{μ}, s^{μ} defined by the matrices R, S are null vectors. ∎

Chapter 8

The Metric and Algebraic Inner Products

In a real finite-dimensional linear metric space, the norm of a vector and the inner product (or scalar product) of two vectors are mutually related.

Given an inner product (u, v), the norm of a vector w is defined as $\|w\|^2 = (w, w)$.

Conversely, given a norm function $\|w\|^2$, an inner product (u, v) is defined by a procedure known as "polarization". This procedure consists of three steps:

(1) Take a linear combination of u and v, the simplest one being

$$w(\lambda) = u + \lambda v,$$

where λ is an auxiliary real parameter.

(2) Compute the norm of w as a second order polynomial in λ. The coefficient of λ is bilinear (linear in u and linear in v). Define *the inner product* (u, v) as one half of this coefficient:

$$
\begin{aligned}
\|w(\lambda)\|^2 &= \|u + \lambda v\|^2 \\
&= \|u\|^2 + \lambda^2 \|v\|^2 + 2\lambda(u, v).
\end{aligned}
\tag{8.1}
$$

(3) Extract (u, v) and eliminate λ. This can be done in three equivalent but computationally different ways.

First solution. By taking the derivative with respect to λ at $\lambda = 0$:

$$(u, v) \stackrel{def}{=} \frac{1}{2} \frac{\partial}{\partial \lambda} \left\| w\left(\lambda\right) \right\|^2 \Big|_{\lambda=0} . \tag{8.2}$$

Second solution. By taking $\lambda = 1$:

$$(u, v) \stackrel{def}{=} \frac{1}{2} \left[\left\| w\left(1\right) \right\|^2 - \left\| u \right\|^2 - \left\| v \right\|^2 \right] . \tag{8.3}$$

Third solution. By taking $\lambda = \pm 1$:

$$(u, v) \stackrel{def}{=} \frac{1}{4} \left[\left\| w\left(1\right) \right\|^2 - \left\| w\left(-1\right) \right\|^2 \right] . \tag{8.4}$$

In the next sections, we shall obtain two types of inner products by applying this general polarization procedure to the metric and algebraic norms of quantions.

8.1   The metric inner product

The substitution of

$$\| Q \|^2 = M\left(Q\right)$$

in the relations (8.2) to (8.4) yields the following general expressions for the *metric inner product*, which we denote by (P, Q) :

$$\left.\begin{array}{l} (P, Q) = \frac{1}{2} \frac{\partial}{\partial \lambda} M\left(P + \lambda Q\right)\big|_{\lambda=0}, \\[2mm] (P, Q) = \frac{1}{2} \left[M\left(P + Q\right) - M\left(P\right) - M\left(Q\right) \right], \\[2mm] (P, Q) = \frac{1}{4} \left[M\left(P + Q\right) - M\left(P - Q\right) \right]. \end{array}\right\} \tag{8.5}$$

By substituting the definition $M\left(Q\right) = Q^{\#}Q$ of the metric norm into any of these equations, one obtains the following convenient working definition of the inner product:

Definition 30 *The **metric inner product** in the algebra of quantions is defined by the relation*

$$(P, Q) = \frac{1}{2} \left(P^{\#}Q + Q^{\#}P \right). \tag{8.6}$$

It is evident that the metric inner product is symmetric,

$$(P,Q) = (Q,P),$$

and that it yields back the metric norm in the case of $P = Q$,

$$(Q,Q) = M(Q).$$

Since the metric norm of a quantion is a complex number, it follows from relation (8.5) that the metric inner product of two quantions is also a complex number.

Substitution of the matrices

$$P = \begin{pmatrix} A & 0 \\ 0 & A \end{pmatrix}, Q = \begin{pmatrix} B & 0 \\ 0 & B \end{pmatrix},$$

$$A = \begin{pmatrix} p_1 & p_3 \\ p_2 & p_4 \end{pmatrix}, B = \begin{pmatrix} q_1 & q_3 \\ q_2 & q_4 \end{pmatrix},$$

into Equation (8.6) yields the explicit expression

$$(P,Q) = \frac{1}{2} (p_1 q_4 - p_2 q_3 - p_3 q_2 + p_4 q_1) I. \tag{8.7}$$

Theorem 31 *A compact form for the metric inner product is*

$$(P,Q) = \frac{i}{2} (p^*| \gamma^2 |q) I. \tag{8.8}$$

Proof. By direct verification. Reminder: $(p^*|$ stands for the q-bra: $(p^*| = \begin{pmatrix} p_1 & p_2 & p_3 & p_4 \end{pmatrix}$. ∎

8.2   The algebraic inner product

The substitution of

$$\|Q\|^2 = A(Q)$$

in the relations (8.2) to (8.4) yields the following general expressions for the *algebraic inner product*, which we denote by $\{P,Q\}$:

$$\left.\begin{aligned}
\{P,Q\} &= \tfrac{1}{2} \tfrac{\partial}{\partial \lambda} A(P + \lambda Q)\big|_{\lambda=0}, \\
\{P,Q\} &= \tfrac{1}{2} [A(P+Q) - A(P) - M(Q)], \\
\{P,Q\} &= \tfrac{1}{4} [A(P+Q) - A(P-Q)].
\end{aligned}\right\} \tag{8.9}$$

Substitution of the definition $A(Q) = Q^\dagger Q$ of the algebraic norm into any of these expressions yields the following convenient working definition of the inner product:

Definition 32 *The **algebraic inner product** in the algebra of quantions is a quantion defined by the relation*

$$\{P, Q\} = \frac{1}{2}\left(P^\dagger Q + Q^\dagger P\right). \tag{8.10}$$

The following properties are evident:

$$\left.\begin{array}{ll} \text{Special case:} & \{Q, Q\} = A(Q) \\ \text{Symmetry:} & \{P, Q\} = \{Q, P\} \\ \text{Hermiticity:} & \{P, Q\}^\dagger = \{P, Q\} \end{array}\right\} \tag{8.11}$$

No simple expression analogous to Theorem 31 exists for the algebraic inner product.

8.3　The classification of quantions

In the matrix representations of complex numbers, quaternions, and quantions, the inverse of an element $Q \neq 0$ is given by the function (6.5), that is

$$Q^{-1} = \frac{1}{m(Q)}Q^\#.$$

Hence, the inverse of a particular quantion Q exists if and only if its metric norm $m(Q)$ is different from zero.

For complex numbers and quaternions, the norms (5.16) and (5.19) are positive definite, so that the inverse exists for every nonvanishing complex number or quaternion.

For quantions, we have

$$m(Q) = \det\begin{pmatrix} q_1 & q_3 \\ q_2 & q_4 \end{pmatrix} = q_1 q_4 - q_2 q_3. \tag{8.12}$$

Hence, the non-trivial solutions of the complex equation

$$q_1 q_4 - q_2 q_3 = 0$$

define a two-dimensional subset of quantions that have no inverse. This suggests the following classification of quantions:

Definition 33 *A quantion Q will be referred to as **singular** or **regular** depending on whether its metric norm $m(Q)$ does or does not vanish.*

To characterize the subset of all singular quantions, it is most convenient to use the standard algebraic concept of an "ideal". Conceptually, ideals are generalizations of the element 0 (zero): In any algebra $\mathcal{A}$, we have the identity

$$(zero) \times (any\ element\ of\ \mathcal{A}) = (zero).$$

More generally, if there exists in $\mathcal{A}$ a subalgebra $\mathcal{J} \subset \mathcal{A}$ such that

$$(any\ element\ of\ \mathcal{J}) \times (any\ element\ of\ \mathcal{A}) = (some\ element\ of\ \mathcal{J}),$$

then $\mathcal{J}$ is called a left ideal in $\mathcal{A}$. In compact form, $\mathcal{JA} = \mathcal{J}$ which generalizes the identity $0 \cdot \mathcal{A} = 0$. If

$$\mathcal{JA} = \mathcal{AJ} = \mathcal{J},$$

$\mathcal{J}$ is a "two-sided ideal".

Using this terminology, the characterization we are seeking is:

Theorem 34 *The subset $\mathcal{J} \subset \mathcal{L}$ of singular quantions is a two-sided maximal ideal in the algebra $\mathcal{L}$ of quantions.*

Proof. We have to verify that $m(PQ) = 0$ if $m(Q) = 0$ or $m(P) = 0$. By Theorem 27, we have

$$m(PQ) = m(P)\, m(Q).$$

Since the metric norm of a quantion is a complex number, it follows that the quantion PQ is singular if and only if P is singular or if Q is singular, which proves the theorem. ∎

The following corollary is of physical significance:

Corollary 35 *The algebraic norm of a regular quantion is a regular quantion. The algebraic norm of a singular quantion is a singular quantion.*

Proof. Take $P = Q^\dagger$ in Theorem 34. ∎

As a help to intuition, Figure 8.1 illustrates the contents of theorems 24 and 35.

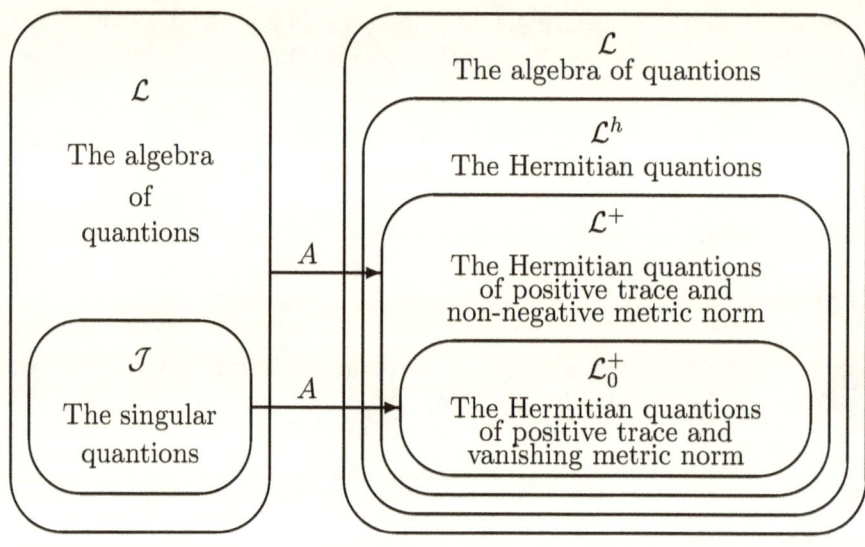

Figure 8.1. The algebraic norm function.

The theorems 24 and 35 are technically trivial but conceptually essential: They provide the structural algebraic foundation for the derivation of the Minkowskian spacetime metric in Chapter 11 and the generalization of Born's interpretation to Zovko's interpretation in Part II.

Chapter 9

The Quantionic Gauge Group

The algebra of quantions has been introduced in its representation by 4×4 block-diagonal matrices. All other representations are mutually related to the block-diagonal form by similarity transformations in the enveloping algebra $\mathcal{M}$,

$$\mathcal{M} \ni M \mapsto SMS^{-1} \in \mathcal{M}, \qquad (9.1)$$

where S is an arbitrary non-singular matrix, $\det S \neq 0$. In other words, all representations are mutually isomorphic. For this reason, we risk no loss of generality by working in the simplest formalism, which is that of block-diagonal matrices.

Some subgroups of similarity transformations are of intrinsic interest. They are the special similarity transformations under which all relevant quantionic concepts remain invariant.

9.1   The invariance groups

The object of this section is to identify the subgroup of similarity transformations (9.1) under which all five products in the table on page 88 are stable. Since the metric and algebraic norms are special cases of the two inner products, it suffices to consider the latter.

Let $\mathcal{F}$ denote a linear mapping $\mathcal{F} : \mathcal{M} \to \mathcal{M}$, and let the symbol $\bigcirc$ stand for any product we may consider. The stability condition for $\bigcirc$ is thus represented by the commutative diagram

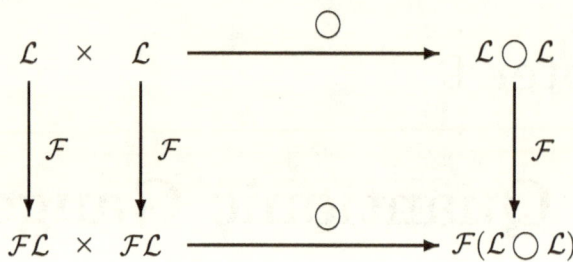

Formally,

$$(\mathcal{F}P) \bigcirc (\mathcal{F}Q) = \mathcal{F}(P \bigcirc Q), \tag{9.2}$$

which, given the operation $\bigcirc$, is to be interpreted as an identity in $P, Q \in \mathcal{L}$ and as a defining equation for the mapping $\mathcal{F}$.

First condition: For the algebraic product itself, meaning that $\bigcirc$ stands for the product of two matrices, $P \bigcirc Q = PQ$, the most general mapping $\mathcal{F}$ is the similarity transformation (9.1), that is,

$$\mathcal{F} : Q \mapsto SQS^{-1} \tag{9.3}$$

The only condition on the matrix S is $\det S \neq 0$. Writing

$$\det S = re^{i\phi},$$

we may take $r = 1$ without loss of generality. On the other hand, we may not take $e^{i\phi} = 1$ because both S^{-1} and the metric norm involve complex conjugation (see the third condition below). Hence, the maximal simplification is

$$\det S = e^{i\phi}. \tag{9.4}$$

Second condition: Applying the transformation (9.3) to the metric inner product, the condition (9.2) reads

$$(P, Q) \mapsto (SPS^{-1}, SQS^{-1}) = S(P, Q)S^{-1}. \tag{9.5}$$

This relation is to be viewed as an identity in P and Q and as a defining equation for S.

Substitution of the definition of the metric inner product into the right-hand side of Equation (9.5) yields

$$r.h.s. = S\left(P,Q\right)S^{-1} = \frac{1}{2}S\left[P^{\#}Q + Q^{\#}P\right]S^{-1}.$$

For the left-hand side, one obtains

$$
\begin{aligned}
l.h.s. \;&=\; \left(SPS^{-1}, SQS^{-1}\right) \\
&=\; \frac{1}{2}\left(\left(SPS^{-1}\right)^{\#}\left(SQS^{-1}\right) + \left(SQS^{-1}\right)^{\#}\left(SPS^{-1}\right)\right) \\
&=\; \frac{1}{2}\left(S^{-1\#}P^{\#}S^{\#}SQS^{-1} + S^{-1\#}Q^{\#}S^{\#}SPS^{-1}\right).
\end{aligned}
$$

From relation (9.4) and the general definition $S^{\#} = \left(\det S\right)S^{-1}$ of the dual follows

$$
\begin{aligned}
S^{\#} \;&=\; e^{i\phi}S^{-1}, \\
S^{-1\#} \;&=\; S^{\#-1} = e^{-i\phi}S, \\
S^{\#}S \;&=\; e^{i\phi}I.
\end{aligned}
$$

Hence

$$
\begin{aligned}
l.h.s. \;&=\; \frac{1}{2}e^{-i\phi}S\left[P^{\#}e^{i\phi}Q + Q^{\#}e^{i\phi}P\right]S^{-1} \\
&=\; \frac{1}{2}S\left[P^{\#}Q + Q^{\#}P\right]S^{-1},
\end{aligned}
$$

which is identical to the right-hand side. The condition (9.5) is thus identically satisfied without imposing any additional conditions on S.

Third condition: Applied to the algebraic inner product, the condition (9.2) reads

$$\{P,Q\} \mapsto \left\{SPS^{-1}, SQS^{-1}\right\} = S\left\{P,Q\right\}S^{-1}, \tag{9.6}$$

This relation is also to be viewed as an identity in P and Q and as a defining equation for S normalized by the condition (9.4).

Substitution of the definition of the algebraic inner product yields the following expressions for the right-hand side

$$r.h.s. = S \{P, Q\} S^{-1} = \frac{1}{2} S \left[P^\dagger Q + Q^\dagger P \right] S^{-1}.$$

For the left-hand side, one obtains

$$
\begin{aligned}
l.h.s. \ &= \ \{SPS^{-1}, SQS^{-1}\} \\
&= \ \frac{1}{2} \left((SPS^{-1})^\dagger (SQS^{-1}) + (SQS^{-1})^\dagger (SPS^{-1}) \right) \\
&= \ \frac{1}{2} \left(S^{-1\dagger} P^\dagger S^\dagger SQS^{-1} + S^{-1\dagger} Q^\dagger S^\dagger SPS^{-1} \right).
\end{aligned}
$$

Hence, the condition $l.h.s. = r.h.s.$ implies

$$S^{-1\dagger} \left[P^\dagger S^\dagger SQ + Q^\dagger S^\dagger SP \right] S^{-1} = S \left[P^\dagger Q + Q^\dagger P \right] S^{-1},$$

which, after multiplication from the left by S^{-1} and from the right by S, simplifies to

$$\left(S^\dagger S \right)^{-1} \left(P^\dagger S^\dagger SQ + Q^\dagger S^\dagger SP \right) = \left(P^\dagger Q + Q^\dagger P \right). \qquad (9.7)$$

Since this relation must be true for all P and all Q in $\mathcal{L}$, let us consider the special case $P = I$. It yields the identity

$$\left(S^\dagger S \right)^{-1} Q^\dagger \left(S^\dagger S \right) = Q^\dagger. \qquad (9.8)$$

which can be rewritten as a vanishing commutator:

$$\left[\left(S^\dagger S \right), \mathcal{L} \right] = 0. \qquad (9.9)$$

One solution is

$$\left(S^\dagger S \right) \in \mathcal{R} \qquad (9.10)$$

because $\mathcal{R}$ is the commutant of $\mathcal{L}$. This implies $S \in \mathcal{R}$, but since

$$\mathcal{L} \ni Q \mapsto SQS^{-1} = QSS^{-1} = Q,$$

the similarity transformation (9.3) with such an S is an identity. The solution (9.10) is thus trivial.

The non-trivial solution of 9.8) is

$$\left(S^\dagger S\right) = I,$$

where $S \notin \mathcal{R}$. It implies that the most general similarity transformation which preserves the metric inner product is given by an arbitrary 4×4 unimodular unitary matrix.

Collecting these results in a theorem, we have

Theorem 36 *The most general group of similarity transformations which commute with the algebraic product of quantions and with the two inner products is the group $SU(4)$ of 4×4 special unitary matrices.*

9.2   The gauge group

We shall now consider the subgroup of the group $SU(4)$ of quantionic automorphisms under which the block-diagonal form of L-type quantions is invariant, that is:

$$\begin{pmatrix} A & 0 \\ 0 & A \end{pmatrix} \mapsto \begin{pmatrix} A' & 0 \\ 0 & A' \end{pmatrix}$$

This group plays an essential role in the algebra of quantions and in its physical applications in Part II, where we recognize it as the gauge group of the electroweak unification. We shall refer to it as the "gauge group", or, more specifically, as the "quantionic gauge group".

Let us write an arbitrary unitary matrix $E \in SU(4)$ in the form

$$E = e^{i\chi} E_0, \tag{9.11}$$

where E_0 is a unitary matrix with unit determinant,

$$\det E_0 = 1.$$

Since $SU(4)$ is a Lie group, we may work with matrices in an infinitesimal neighborhood of the unit matrix without loss of generality. In block form, such a matrix is given by the expression

$$E_0 = I + i\varepsilon \begin{pmatrix} H & M \\ M^\dagger & K \end{pmatrix}.$$

The matrices H and K are arbitrary 2×2 Hermitian matrices, while M is an arbitrary complex 2×2 matrix. Hence, the inverse E_0^{-1} is

$$E_0^{\dagger} = I - i\varepsilon \begin{pmatrix} H & M \\ M^{\dagger} & K \end{pmatrix}.$$

The similarity transformation of an arbitrary quantion Q is thus:

$$Q \mapsto SQS^{-1} = EQE^{\dagger} = E_0 Q E_0^{\dagger} = Q + i\varepsilon Z, \qquad (9.12)$$

where Z represents the commutator

$$\begin{aligned}
Z &= \begin{pmatrix} H & M \\ M^{\dagger} & K \end{pmatrix} \begin{pmatrix} A & 0 \\ 0 & A \end{pmatrix} - \begin{pmatrix} A & 0 \\ 0 & A \end{pmatrix} \begin{pmatrix} H & M \\ M^{\dagger} & K \end{pmatrix} \\
&= \begin{pmatrix} [H,A] & [M,A] \\ [M^{\dagger},A] & [K,A] \end{pmatrix}.
\end{aligned}$$

For the transformation (9.12) to preserve the block-diagonal form of Q, the commutator Z must be an L-type quantion for all A. This requirement implies

$$\begin{aligned}
[M, A] &= 0, \\
[H, A] &= [K, A].
\end{aligned}$$

Hence,

$$\begin{aligned}
M &= 0, \\
K &= H.
\end{aligned}$$

By integration (Stone's theorem), we get the finite transformation

$$E_0 = \exp i \begin{pmatrix} H & 0 \\ 0 & H \end{pmatrix} = \begin{pmatrix} U_0 & 0 \\ 0 & U_0 \end{pmatrix},$$

where U_0 is an arbitrary 2×2 unitary unimodular matrix, $\det U_0 = 1$.

Definition 37 *The group of transformations*

$$E = e^{i\chi} E_0 = e^{i\chi} \begin{pmatrix} U_0 & 0 \\ 0 & U_0 \end{pmatrix}, \qquad (9.13)$$

*where U_0 is an arbitrary 2×2 unimodular unitary matrix, will be denoted by $U_q(1)$ and referred to as the **quantionic gauge group**.*

Since $e^{i\chi} \in U(1)$ and $U_0 \in SU(2)$, the group $U_q(1)$ is isomorphic with the group $SU(2) \times U(1)$.

Knowing that the electroweak gauge group of the Standard model is also $SU(2) \times U(1)$, one may wonder whether this coincidence is spurious or profound. We shall see in Part II that $U_q(1)$ classifies the four arbitrary vector potentials that appear in the derivation of the quantionic form of Dirac's equation — which strongly suggests that the coincidence is not spurious.

The complex phase transformation

$$\psi \longmapsto \psi' = e^{i\chi}\psi \tag{9.14}$$

is the most general transformation which preserves the algebraic norm of complex numbers:

$$\psi^*\psi \longmapsto \psi'^*\psi' = \left(e^{i\chi}\psi\right)^* \left(e^{i\chi}\psi\right) = \psi^*\psi.$$

Similarly, the quantionic phase transformation

$$Q \longmapsto Q' = EQ \tag{9.15}$$

is the most general transformation which preserves the algebraic norm of quantions:

$$Q^\dagger Q \longmapsto Q'^\dagger Q' = (EQ)^\dagger (EQ) = Q^\dagger E^\dagger E Q = Q^\dagger Q.$$

The quantionic gauge group $U_q(1)$ is thus the generalization to quantions of the complex gauge group $U(1)$. These gauge groups are special cases, for $n = 1$, of the unitary group $SU(n)$ in a complex n-dimensional Hilbert space, and of the analogous group $SU_q(n)$ in a quantionic n-dimensional Hilbert space, as defined in Chapter 12.

By analogy with the "complex phase factor" $e^{i\chi}$, the unit quantion

$$E_0 = \begin{pmatrix} (E_0)_{red} & 0 \\ 0 & (E_0)_{red} \end{pmatrix}$$

will be referred to as the "pure quantionic phase factor".

In the next section we discuss several ways of writing the quantionic phase factor.

9.3   The various expressions for gauge transformations

(1) The matrix form

In terms of three compact parameters,[1] α, β, γ, a general expression for the matrix E_{red} is

$$E_{red} = e^{i\chi}U = e^{i\chi} \begin{pmatrix} \cos\gamma e^{i\alpha} & -\sin\gamma e^{i\beta} \\ \sin\gamma e^{-i\beta} & \cos\gamma e^{-i\alpha} \end{pmatrix}, \tag{9.16}$$

where U is a shorter notation for $(E_0)_{red}$. In the non-reduced form:

$$E = e^{i\chi}E_0 = e^{i\chi} \begin{pmatrix} U & 0 \\ 0 & U \end{pmatrix}$$

$$= e^{i\chi} \begin{pmatrix} \cos\gamma e^{i\alpha} & -\sin\gamma e^{i\beta} & 0 & 0 \\ \sin\gamma e^{-i\beta} & \cos\gamma e^{-i\alpha} & 0 & 0 \\ 0 & 0 & \cos\gamma e^{i\alpha} & -\sin\gamma e^{i\beta} \\ 0 & 0 & \sin\gamma e^{-i\beta} & \cos\gamma e^{-i\alpha} \end{pmatrix}. \tag{9.17}$$

(2) The exponential form

In term of an arbitrary unit vector $\vec{m}$, that is, $\vec{m} \cdot \vec{m} = 1$, we may write

$$E = e^{i\chi} \exp\left(i\phi\vec{m} \cdot \vec{\Lambda}\right). \tag{9.18}$$

Unitarity is readily verified: The expressions

$$E_0 = \exp\left(i\phi\vec{m} \cdot \vec{\Lambda}\right) = \cos\phi I + i\sin\phi \, \vec{m} \cdot \vec{\Lambda}$$

$$E_0^\dagger = \exp\left(-i\phi\vec{m} \cdot \vec{\Lambda}\right) = \cos\phi I - i\sin\phi \, \vec{m} \cdot \vec{\Lambda}$$

imply

$$E_0^\dagger E_0 = \left(\cos^2\phi + \sin^2\phi\right) I = I.$$

The rather complicated functional relationships between the parameters α, β, γ and m_1, m_2, m_3 are derived in [10]. Since it appears that they will not be needed, we may ignore them.

---

[1] Readers who might be comparing this matrix E_0 with the corresponding matrix in the book [10] will notice a difference in the sign of $\sin\gamma$. The present choice is preferable because, in the case of $\alpha = \beta = 0$, the matrix E_0 assumes the form of a counter-clockwise rotation, which is the standard convention.

(3) A factorization of gauge transformations

Consider the general gauge transformation operator

$$E = e^{i\chi} E_0 \left(\gamma, \alpha, \beta \right) = e^{i\chi} \begin{pmatrix} U \left(\gamma, \alpha, \beta \right) & 0 \\ 0 & U \left(\gamma, \alpha, \beta \right) \end{pmatrix}, \qquad (9.19)$$

where U is defined by (9.16), i.e.,

$$U \left(\gamma, \alpha, \beta \right) = \begin{pmatrix} \cos \gamma e^{i\alpha} & -\sin \gamma e^{i\beta} \\ \sin \gamma e^{-i\beta} & \cos \gamma e^{-i\alpha} \end{pmatrix}. \qquad (9.20)$$

The matrix U can be written as a product of the following one-parametric matrices:

$$\left. \begin{array}{l} R \left(\gamma \right) \overset{def}{=} U \left(\gamma, 0, 0 \right) = \begin{pmatrix} \cos \gamma & -\sin \gamma \\ \sin \gamma & \cos \gamma \end{pmatrix}, \\[2mm] A \left(\alpha \right) \overset{def}{=} U \left(0, \alpha, 0 \right) = \begin{pmatrix} e^{i\alpha} & 0 \\ 0 & e^{-i\alpha} \end{pmatrix}, \\[2mm] B \left(\beta \right) \overset{def}{=} U \left(0, 0, \beta \right) = \begin{pmatrix} 0 & ie^{i\beta} \\ -ie^{-i\beta} & 0 \end{pmatrix}. \end{array} \right\} \qquad (9.21)$$

The transformations B do not form a group, but the transformations R and A form two one-parametric groups and satisfy the following identities:

$$\left. \begin{array}{l} B \left(\beta \right) B \left(\beta \right) = I, \\ B \left(\beta \right) B \left(-\beta \right) = A \left(2\beta \right), \\ B \left(-\beta \right) B \left(\beta \right) = A \left(-2\beta \right), \\ A \left(\alpha \right) B \left(\beta \right) = B \left(\beta + \alpha \right), \\ B \left(\beta \right) A \left(\alpha \right) = B \left(\beta - \alpha \right). \end{array} \right\} \qquad (9.22)$$

Theorem 38 *The general unitary transformation $E_0 \left(\gamma, \alpha, \beta \right)$ admits a unique factorization which may be written in terms of the component block matrices U,*

$$U \left(\gamma, \alpha, \beta \right) = A \left(\frac{\alpha}{2} \right) B \left(\frac{\beta}{2} \right) R \left(\gamma \right) B \left(\frac{\beta}{2} \right) A \left(\frac{\alpha}{2} \right), \qquad (9.23)$$

or, equivalently

$$U \left(\gamma, \alpha, \beta \right) = B \left(\frac{\beta}{2} \right) A \left(\frac{\alpha}{2} \right) R \left(\gamma \right) A \left(\frac{\alpha}{2} \right) B \left(\frac{\beta}{2} \right). \qquad (9.24)$$

Proof. We first compute

$$U\left(\gamma,0,\beta\right) \;=\; B\left(\frac{\beta}{2}\right) R\left(\gamma\right) B\left(\frac{\beta}{2}\right)$$

$$= \begin{pmatrix} 0 & ie^{i\frac{\beta}{2}} \\ -ie^{-i\frac{\beta}{2}} & 0 \end{pmatrix} \begin{pmatrix} \cos\gamma & -\sin\gamma \\ \sin\gamma & \cos\gamma \end{pmatrix} \begin{pmatrix} 0 & ie^{i\frac{\beta}{2}} \\ -ie^{-i\frac{\beta}{2}} & 0 \end{pmatrix}$$

$$= \begin{pmatrix} \cos\gamma & -e^{i\beta}\sin\gamma \\ e^{-i\beta}\sin\gamma & \cos\gamma \end{pmatrix},$$

and then

$$U\left(\gamma,\alpha,\beta\right) = A\left(\frac{\alpha}{2}\right) U\left(\gamma,0,\beta\right) A\left(\frac{\alpha}{2}\right)$$

$$= \begin{pmatrix} e^{i\frac{\alpha}{2}} & 0 \\ 0 & e^{-i\frac{\alpha}{2}} \end{pmatrix} \begin{pmatrix} \cos\gamma & -e^{i\beta}\sin\gamma \\ e^{-i\beta}\sin\gamma & \cos\gamma \end{pmatrix} \begin{pmatrix} e^{i\frac{\alpha}{2}} & 0 \\ 0 & e^{-i\frac{\alpha}{2}} \end{pmatrix}$$

$$= \begin{pmatrix} e^{i\alpha}\cos\gamma & -e^{i\beta}\sin\gamma \\ e^{-i\beta}\sin\gamma & e^{-i\alpha}\cos\gamma \end{pmatrix}.$$

This completes the proof. ∎

Theorem 39 *The matrices $U\left(\gamma,0,\beta\right)$ form a group only if the parameter β is fixed.*

Proof. One verifies that the product

$$U\left(\gamma,0,\beta\right) U\left(\gamma',0,\beta'\right)$$

$$= \begin{pmatrix} \cos\gamma & -e^{i\beta}\sin\gamma \\ e^{-i\beta}\sin\gamma & \cos\gamma \end{pmatrix} \begin{pmatrix} \cos\gamma' & -e^{i\beta'}\sin\gamma' \\ e^{-i\beta'}\sin\gamma' & \cos\gamma' \end{pmatrix}$$

is of the same type as the factors only if $\beta' = \beta$, in which case

$$U\left(\gamma,0,\beta\right) U\left(\gamma',0,\beta\right)$$

$$= \begin{pmatrix} \cos\left(\gamma+\gamma'\right) & -e^{i\beta}\sin\left(\gamma+\gamma'\right) \\ e^{-i\beta}\sin\left(\gamma+\gamma'\right) & \cos\left(\gamma+\gamma'\right) \end{pmatrix}$$

$$= U\left(\gamma+\gamma',0,\beta\right),$$

completing the proof. ∎

Chapter 10

The Polar Structure of Quantions

In Table 1.1 on page 14, the entry "polar structure" in the second row from the bottom refers to the Eulerian factorization. For a complex number z, it is the factorization

$$z = e^{i\phi} \, r \qquad (10.1)$$

into a modulus $r \in \mathbb{R}^+$ and a phase factor $e^{i\phi} \in U(1)$.

This factorization is symmetric (one phase factor and one real modulus), while the factorization of quaternions, $q = e^{i\phi}e^{j\phi}e^{k\phi} \, r$, is not (three non-commuting phase factors and one real modulus).

Due to their complex structure (see Table 1.1), quantions admit a symmetric polar factorization

$$Q = ER, \qquad (10.2)$$

which is conceptually identical to (10.1). The analogies are

$$\left. \begin{array}{l} U(1) \ni e^{i\phi} \rightleftarrows E \in U_q(1), \\ \mathbb{R}^+ \ni r \rightleftarrows R \in \mathcal{L}^+. \end{array} \right\} \qquad (10.3)$$

While ϕ and r are real parameters, accounting for the two dimensions of the complex plane, both E and R are four-parametric objects.

We refer to R and E as the **quantionic modulus** and **quantionic phase factor** respectively

The ordering ER of the non-commuting factors will be considered as the 'reference ordering'.

Since the phase factor E may be viewed as a quantionic gauge transformation, $E \in U_q(1)$, we have the following theorem:

Theorem 40 *The convex space $\mathcal{L}^+$ of positive Hermitian quantions is invariant under the group of transformations $U_q(1)$.*

Proof. For $E \in U_q(1)$ and $R \in \mathcal{L}^+$, the transformed quantion

$$R' = ERE^\dagger$$

has the following properties:

(a) It is Hermitian:

$$R'^\dagger = \left(ERE^\dagger\right)^\dagger = E^{\dagger\dagger} R^\dagger E^\dagger = ERE^\dagger = R'.$$

(b) It has the same determinant as R :

$$\det R' = \det\left(ERE^\dagger\right) = \left(\det EE^\dagger\right)(\det R) = \det R.$$

(c) It has the same trace as R :

$$Tr\left(R'\right) = Tr\left(ERE^\dagger\right) = Tr\left(E^\dagger ER\right) = Tr\left(R\right).$$

Thus, $\mathcal{L}^+$ and $\mathcal{L}_0^+$ are mapped onto themselves. ■

The next task is to extend to quantions the solutions

$$\left.\begin{array}{l} r = +\sqrt{z^*z}, \\ e^{i\phi} = z/r, \end{array}\right\} \tag{10.4}$$

and to investigate the factors R and E.

10.1   The quantionic modulus

Substitution of relation (10.2) into the algebraic norm yields

$$A(Q) = Q^\dagger Q = (ER)^\dagger (ER) = R^\dagger E^\dagger ER = R^2. \tag{10.5}$$

Hence, formally:
$$R = \sqrt{A(Q)} = \sqrt{Q^\dagger Q}. \tag{10.6}$$

As shown in the next theorem, a uniquely distinguished solution always exists for the square root.

Theorem 41 *Given an arbitrary quantion $Q \in \mathcal{L}$, the modulus R is uniquely defined. If Q is regular, $R \in \mathcal{L}^+$; if Q is singular, $R \in \mathcal{L}_0^+$.*

Proof. Since $A(Q)$ is a positive Hermitian quantion,

$$A(Q) = \begin{pmatrix} Q_{red}^\dagger Q_{red} & 0 \\ 0 & Q_{red}^\dagger Q_{red} \end{pmatrix} \in \mathcal{L}^+,$$

the reduced matrix may be diagonalized:

$$Q_{red}^\dagger Q_{red} = \begin{pmatrix} \lambda_1 & 0 \\ 0 & \lambda_2 \end{pmatrix}.$$

Consequently,

$$Tr\left(Q_{red}^\dagger Q_{red}\right) \geqslant 0 \text{ implies } \lambda_1 + \lambda_2 \geqslant 0,$$
$$\det\left(Q_{red}^\dagger Q_{red}\right) \geqslant 0 \text{ implies } \lambda_1 \lambda_2 \geqslant 0.$$

These two conditions imply $\lambda_1 \geqslant 0$ and $\lambda_2 \geqslant 0$. Clearly $\lambda_1 = \lambda_2 = 0$ if and only if $Q = 0$.

In terms of eigenvalues, there are four solutions for R. To list them, let μ_1, μ_2 denote the positive square roots of λ_1, λ_2 :

$$\mu_1 \overset{def}{=} \sqrt{\lambda_1},$$
$$\mu_2 \overset{def}{=} \sqrt{\lambda_2}.$$

The diagonalized solutions for R are thus:

$$R = \begin{pmatrix} \mu_1 & 0 \\ 0 & \mu_2 \end{pmatrix}, \qquad R_2 = -R,$$

$$R_1 = \begin{pmatrix} -\mu_1 & 0 \\ 0 & \mu_2 \end{pmatrix}, \qquad R_3 = -R_1.$$

The solutions R_2 and R_3 are redundant because the minus sign can be absorbed by the phase factor. Similarly, R_1 is redundant because the modification

$$E \to F = E \begin{pmatrix} -1 & 0 \\ 0 & 1 \end{pmatrix},$$

of the phase factor E yields

$$Q = E \begin{pmatrix} -\mu_1 & 0 \\ 0 & \mu_2 \end{pmatrix} = F \begin{pmatrix} \mu_1 & 0 \\ 0 & \mu_2 \end{pmatrix}.$$

The solution R, which always exists, is thus the only solution.   ■

Computing $R = \sqrt{Q^\dagger Q}$ may be either very difficult or trivial, depending on the selected parametrization of the quantion Q.

Theorem 42 *The parametrization of R in which it is evident that $Q^\dagger Q$ is in $\mathcal{L}^+$ for a regular quantion is*

$$R = a \left(\cosh \omega \; I + \sinh \omega \; \vec{n} \cdot \vec{\Lambda} \right), \tag{10.7}$$

where a is any positive number and $\vec{n}$ is an arbitrary unit vector, $\vec{n} \cdot \vec{n} = 1$. Similarly, the parametrization in which it is evident that $Q^\dagger Q$ is in $\mathcal{L}_0^+$ for a singular quantion is

$$R = a \left(I + \vec{n} \cdot \vec{\Lambda} \right). \tag{10.8}$$

Proof. For a regular quantion, the expression (10.7) yields

$$Q^\dagger Q = R^2 = a^2 \left(\cosh 2\omega \; I + \sinh 2\omega \; \vec{n} \cdot \vec{\Lambda} \right). \tag{10.9}$$

The trace

$$Tr \left(Q^\dagger Q \right) = 4a^2 \cosh 2\omega \tag{10.10}$$

is always positive.

The determinant of the reduced matrix,

$$\det \left(Q^\dagger_{red} Q_{red} \right) = a^4 \left(\cosh^2 2\omega - \sinh^2 2\omega \right) = a^4, \tag{10.11}$$

is also positive.

For singular quantions, the parametrization (10.8) yields

$$Q^\dagger Q = 2a^2 \left(I + \vec{n} \cdot \vec{\Lambda} \right). \tag{10.12}$$

The trace

$$Tr \left(Q^\dagger Q \right) = 8a^2$$

is always positive.

The determinant of the reduced matrix,

$$\det \left(Q^\dagger_{red} Q_{red} \right) = 4a^4 \left(I - \vec{n} \cdot \vec{n} I \right) \equiv 0, \tag{10.13}$$

vanishes identically. ■

This completes the analysis of the properties of the modulus.

10.2   The phase factors

We observe that Equation (10.2) formally yields

$$E = QR^{-1}, \tag{10.14}$$

but this is meaningful only if $m(R) \neq 0$. Thus, regular and singular quantions have to be considered separately.

Regular quantions

If Q is regular, so is R, in which case R^{-1} exists:

$$R^{-1} = \frac{1}{m(R)} R^\# = \frac{1}{\det \left(Q^\dagger_{red} Q_{red} \right)} R^\#. \tag{10.15}$$

The phase factor is then given by relation (10.14):

$$E = \frac{1}{\det \left(Q^\dagger_{red} Q_{red} \right)} Q R^\#. \tag{10.16}$$

In the parametric solution (10.7), we have

$$R^{-1} = a^{-1} \left(\cosh \omega \, I - \sinh \omega \, \vec{n} \cdot \vec{\Lambda} \right), \tag{10.17}$$

and hence, by relation (10.14):

$$E = a^{-1}Q\left(\cosh\omega\, I - \sinh\omega\, \vec{n}\cdot\vec{\Lambda}\right). \tag{10.18}$$

This completes the factorization of regular quantions.

Singular quantions

For singular quantions, the phase factor is not defined by relation (10.14) because $\det R = 0$.

Theorem 43 *In the polar factorization of a singular quantion, the phase factor E is defined up to a one-parametric group of transformations.*

Proof. Let us write the modulus R in the diagonalized form

$$R = \begin{pmatrix} S & 0 \\ 0 & S \end{pmatrix},$$

where

$$S = \begin{pmatrix} \lambda_1 & 0 \\ 0 & \lambda_2 \end{pmatrix}.$$

Taking for the phase factor E the expressions obtained in Chapter 9,

$$E = e^{i\chi}\begin{pmatrix} U & 0 \\ 0 & U \end{pmatrix},$$

$$U = \begin{pmatrix} \cos\gamma e^{i\alpha} & -\sin\gamma e^{i\beta} \\ \sin\gamma e^{-i\beta} & \cos\gamma e^{-i\alpha} \end{pmatrix},$$

the reduced matrix Q_{red} reads

$$\begin{aligned}
Q_{red} &= e^{i\chi}\begin{pmatrix} \cos\gamma e^{i\alpha} & -\sin\gamma e^{i\beta} \\ \sin\gamma e^{-i\beta} & \cos\gamma e^{-i\alpha} \end{pmatrix}\begin{pmatrix} \lambda_1 & 0 \\ 0 & \lambda_2 \end{pmatrix} \\
&= \begin{pmatrix} \lambda_1 e^{i(\alpha+\chi)}\cos\gamma & -\lambda_2 e^{i(\beta+\chi)}\sin\gamma \\ \lambda_1 e^{-i(\beta-\chi)}\sin\gamma & \lambda_2 e^{-i(\alpha-\chi)}\cos\gamma \end{pmatrix}.
\end{aligned}$$

If Q is a regular quantion, the parameters α and β appear with plus and minus signs, so that no transformation of these parameters leaves Q invariant.

If Q is singular due to $\lambda_2 = 0$, then

$$Q = \begin{pmatrix} \lambda_1 e^{i(\alpha+\chi)} \cos\gamma & 0 \\ \lambda_1 e^{-i(\beta-\chi)} \sin\gamma & 0 \end{pmatrix} \qquad (10.19)$$

is invariant under the group of transformations

$$\begin{pmatrix} \alpha \\ \beta \\ \chi \end{pmatrix} \mapsto \begin{pmatrix} \alpha - \xi \\ \beta + \xi \\ \chi + \xi \end{pmatrix} \qquad (10.20)$$

for an arbitrary parameter ξ.

Similarly, if Q is singular due to $\lambda_1 = 0$, we have

$$Q = \begin{pmatrix} 0 & -\lambda_2 e^{i(\beta+\chi)} \sin\gamma \\ 0 & \lambda_2 e^{-i(\alpha-\chi)} \cos\gamma \end{pmatrix}, \qquad (10.21)$$

and the transformations that leave this matrix invariant are

$$\begin{pmatrix} \alpha \\ \beta \\ \chi \end{pmatrix} \mapsto \begin{pmatrix} \alpha + \xi \\ \beta - \xi \\ \chi + \xi \end{pmatrix}. \qquad (10.22)$$

In both cases, the parameter $\xi \in [0, 2\pi)$ is arbitrary. ∎

10.3   The reordering of factors

For complex numbers, the modulus and phase factor commute. For general quantions, they don't:

$$ER \neq RE.$$

One of the reasons the ordering 'phase factor times modulus' was selected as the reference ordering for computations is that the expression it yields for the algebraic norm does not contain the phase factor:

$$A(Q) = Q^\dagger Q = (ER)^\dagger (ER) = R^\dagger E^\dagger E R = R^2.$$

The phase factor does appear in the opposite ordering,

$$A(Q) = Q^\dagger Q = (RE)^\dagger (RE) = E^\dagger R^\dagger R E = E^\dagger R^2 E,$$

but it does not affect Hermiticity:

$$\left(E^{\dagger} R^2 E\right)^{\dagger} = E^{\dagger} R^2 E^{\dagger\dagger} = E^{\dagger} R^2 E.$$

The relationship between the two factorizations is given by the following theorem:

Theorem 44 *For a regular quantion Q, reversing the order of factors does not affect the phase factor, but it transforms the modulus:*

$$ER = Q = SE, \tag{10.23}$$

where

$$S = ERE^{\dagger}. \tag{10.24}$$

Proof. Multiply (10.23) by E from the right. ∎

The singular decomposition

Theorem 45 *An arbitrary regular quantion admits a unique decomposition into a sum of two singular quantions, $Q = Q_1 + Q_2$.*

Proof. Let us write Q in the form

$$Q = ER = EU \begin{pmatrix} \lambda_1 & 0 & 0 & 0 \\ 0 & \lambda_2 & 0 & 0 \\ 0 & 0 & \lambda_1 & 0 \\ 0 & 0 & 0 & \lambda_2 \end{pmatrix} U^{\dagger},$$

where U is the matrix that diagonalizes R. If either of the two eigenvalues vanishes, Q is singular.

Let us now define Q_1 and Q_2 as

$$Q_1 = ER_1 = EU \begin{pmatrix} \lambda_1 & 0 & 0 & 0 \\ 0 & 0 & 0 & 0 \\ 0 & 0 & \lambda_1 & 0 \\ 0 & 0 & 0 & 0 \end{pmatrix} U^{\dagger},$$

$$Q_2 = ER_2 = EU \begin{pmatrix} 0 & 0 & 0 & 0 \\ 0 & \lambda_2 & 0 & 0 \\ 0 & 0 & 0 & 0 \\ 0 & 0 & 0 & \lambda_2 \end{pmatrix} U^{\dagger}.$$

It is now evident that the singular decomposition $Q = Q_1 + Q_2$ is always possible and that it is unique. ∎

The separation of the trace

An arbitrary quantion may be written in the form

$$Q = zI + Q_0,$$

where Q_0 is an arbitrary complex traceless 4×4 matrix. For the modulus R, the corresponding expression is

$$R = rI + R_0,$$

where r is a positive real number and R_0 a traceless Hermitian matrix:

$$R_0 = \begin{pmatrix} s & z^* & 0 & 0 \\ z & -s & 0 & 0 \\ 0 & 0 & s & z^* \\ 0 & 0 & z & -s \end{pmatrix}.$$

We can now determine the geometric meaning of the transformation (10.24):

Theorem 46 *The unitary transformation $S = ERE^\dagger$ of the modulus of a quantion is a rotation in the Euclidian 3-space of traceless Hermitian quantions.*

Proof. By relation (7.9), the metric norm is invariant under the transformation (10.24):

$$\begin{aligned} M\left(S\right) &= M\left(E\right) M\left(R\right) M\left(E^\dagger\right) = M\left(E\right) M\left(E^\dagger\right) M\left(R\right) \\ &= M\left(EE^\dagger\right) M\left(R\right) = M\left(I\right) M\left(R\right) = M\left(R\right). \end{aligned}$$

To show that the trace is also invariant, let us separate out the traceless part R_0:

$$S = sI + S_0 = ERE^\dagger = E_0\left(rI + R_0\right) E_0^\dagger = rI + E_0 R_0 E_0^\dagger.$$

Then,

$$Tr\,(S) \;=\; Tr\,(R)\,,$$
$$S_0 \;=\; E_0 R_0 E_0^\dagger.$$

Hence, the scalar component $Tr\,(R)$ is invariant.

The transformation which keeps this component invariant and transforms the three-dimensional space of traceless matrices R_0 into itself in such a way that the metric norm $M\,(R_0)$ remains invariant is formally a 3-rotation. This proves the assertion. ∎

To obtain the explicit form of the rotation, let us write

$$R_0 \;=\; \vec{r}_0 \cdot \vec{\Lambda},$$
$$E_0 \;=\; \exp\left(i\vec{m} \cdot \vec{\Lambda}\right).$$

Then,

$$S_0 = E_0 R_0 E_0^\dagger = \exp\left(i\vec{m} \cdot \vec{\Lambda}\right)\left(\vec{r}_0 \cdot \vec{\Lambda}\right)\exp\left(-i\vec{m} \cdot \vec{\Lambda}\right)$$

For an infinitesimal transformation, $\vec{m}$ is an infinitesimal vector. To make this point explicit, let us write $\varepsilon\vec{m}$ instead of $\vec{m}$, so that $\vec{m}$ can be normalized for convenience: $\vec{m} \cdot \vec{m} = 1$. Then, to first order in the infinitesimal parameter ε, one obtains

$$\begin{aligned}
S_0 \;&=\; \left(I + i\varepsilon\vec{m} \cdot \vec{\Lambda}\right)\left(\vec{r}_0 \cdot \vec{\Lambda}\right)\left(I - i\varepsilon\vec{m} \cdot \vec{\Lambda}\right) \\
&=\; \vec{r}_0 \cdot \vec{\Lambda} + i\varepsilon\left[\vec{m} \cdot \vec{\Lambda}, \vec{r}_0 \cdot \vec{\Lambda}\right] \\
&=\; [\vec{r}_0 + 2\varepsilon\,(\vec{m} \times \vec{r}_0)] \cdot \vec{\Lambda}.
\end{aligned}$$

This means that the vector $\vec{r}_0$ transforms by the rule

$$\vec{r}_0 \mapsto \vec{r}_0 + 2\varepsilon\vec{m} \times \vec{r}_0, \qquad (10.25)$$

which represents a counter-clockwise rotation around the axis defined by $\vec{m}$. For a finite parameter ϕ, we thus have:

$$s_0 = R_{(\vec{m},2\phi)}\vec{r}_0, \qquad (10.26)$$

where $R_{(\vec{m},\xi)}$ is the operator of counter-clockwise rotation by a angle ξ about the axis defined by the unit vector $\vec{m}$.

Chapter 11

The Geometry of Quantions

Just as the algebra of quantions is an extension of the algebra of complex numbers, the geometry of quantions is an extension of the geometry of complex numbers in the complex plane. While technically very simple, this extension is conceptually crucial because the counterpart of the complex plane is a linear space with Minkowski metric. This introduces the relativistic spacetime as a theorem.

Since the present chapter is at the transition from the algebra to the geometry of quantions, this may be a good place to put in perspective the most important concepts of the mathematics of quantions developed so far. These concepts are illustrated in Figure 11.1. The mathematical or physical interpretations of these objects are indicated in the large ovals. The two ovals at the bottom are emphasized for representing standard physics, namely nonrelativistic quantum mechanics and relativity.

Let us begin a walk through this graph at the doubly-framed full algebra $\mathcal{M}$ of complex 4×4 matrices:

(1) We arbitrarily select in $\mathcal{M}$ any two maximal mutual commutants, referred to as the subalgebras $\mathcal{L}$ and $\mathcal{R}$. These subalgebras are related by a similarity transformation (the matrix W) which makes them intrinsically indistinguishable.

(2) We take one of the subalgebras — we have selected $\mathcal{L}$ — as the

definition of the algebra of quantions. For convenience, the matrices in $\mathcal{L}$ are taken to be in block-diagonal form. Equivalent representations are related by similarity transformations.

(3) The second subalgebra, $\mathcal{R}$, is algebraically redundant for being related to $\mathcal{L}$ by the specific similarity transformation W. But we shall see in Chapter 13 that it plays a key role in quantionic analysis: It is essential to the existence of a derivation operator $\mathcal{D}$ in $\mathcal{L}$, as indicated by the caption "Differential structure".

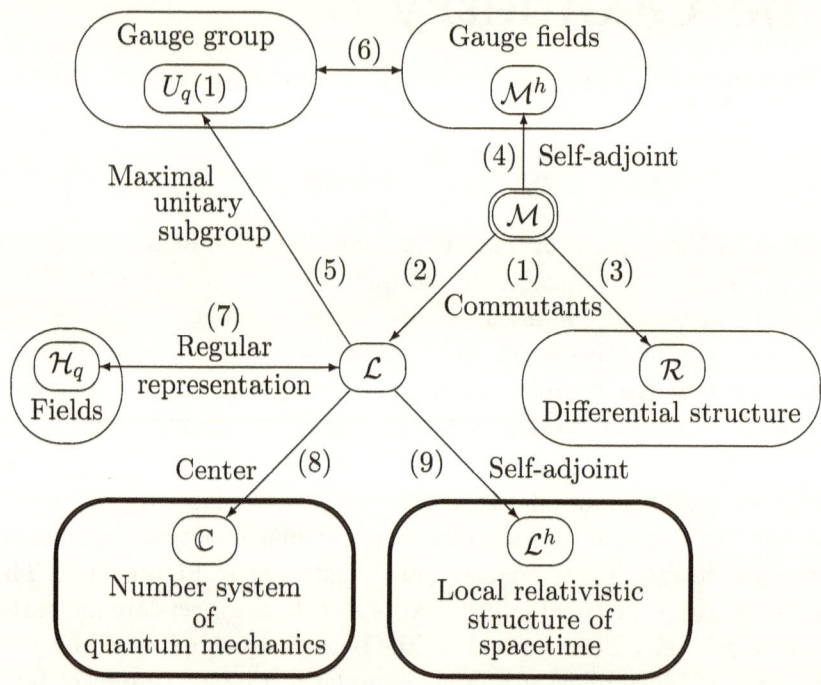

Figure 11.1: The structuring of the algebra $\mathcal{M}$.

(4) Consider the sixteen-dimensional linear subspace $\mathcal{M}^h \subset \mathcal{M}$ of Hermitian matrices. It was shown in Theorem 19 that $\mathcal{M}^h$ naturally decomposes into four real vector fields in Minkowski space. As briefly indicated in the discussion on page 71 and proved exactly in Part II, these vectors may be interpreted as external potentials or as differential connections in the quantionic field equation (see (6) below).

(5) By analogy with the gauge group $U(1)$ in the complex domain, the gauge group $U_q(1)$ of the algebra $\mathcal{L}$ of quantions is defined as the group of transformations that preserve the algebraic norm of quantions. Equivalently, it is the maximal unitary group contained in the algebra $\mathcal{L}$.

(6) As shown in Part II, the four vector fields equivalent to the Hermitian matrices in $\mathcal{M}^h$ are related to the four parameters of the quantionic gauge group $U_q(1)$ as differential connections if the quantionic derivation operator $\mathcal{D}$ is made covariant with respect to this group.

(7) The very special properties of the algebra $\mathcal{L}$ are such that, as a linear space, $\mathcal{L}$ is isomorphic with its four-dimensional regular representation space $\mathcal{H}_q$. The difference is that $\mathcal{H}_q$ has a Hilbert space structure while $\mathcal{L}$ has an algebraic structure.

(8) The center of $\mathcal{L}$ is the field $\mathbb{C}$ of complex numbers. One thus retrieves nonrelativistic quantum mechanics by restricting the algebra of quantions to its center.

(9) The restriction of the algebra $\mathcal{L}$ of complex quantions to its real linear subspace $\mathcal{L}^h$ of Hermitian quantions leads to relativity. The reason is that the subspace $\mathcal{L}^h$ is metrically isomorphic with the linear Minkowski space M_0^4. The present chapter is dedicated to this result.

The two bottom branches in Figure 11.1 show how the double structure of the algebra $\mathcal{L}$ of quantions gives rise to the structural unification of relativity and quantum mechanics: The algebraic structure of $\mathcal{L}$ has the properties needed to support quantum mechanics (in particular, nonrelativistic quantum mechanics in the limit $\mathcal{L} \to \mathbb{C}$), while the real geometric structure of $\mathcal{L}^h$ *is* the local relativistic structure of spacetime.

11.1 The quantionic Gaussian space

The "metric norm" $m(Q) \in \mathbb{C}$ and the "metric dual" $Q^\# \in \mathcal{L}$ are defined respectively by the relations

$$m(Q) = \det(Q_{red}), \qquad (11.1)$$
$$Q^\# = m(Q)\, Q^{-1}. \qquad (11.2)$$

These two norms have been introduced in Chapter 6 as purely algebraic concepts.

The 'geometry of quantions' consists in rewriting these algebraic concepts and related theorems in terms of linear metric space concepts. This introduces no extraneous assumptions or concepts whose definitions could not be traced to Definition 1 on page 27.

The underlying idea stems from the geometric representation of complex numbers: Complex numbers $z = x + iy$ are plotted as points in a plane (the "complex plane", or "Gaussian plane"), and the arithmetic operations with these numbers are reinterpreted as geometric constructs in the this plane.

As is evident from the following timetable of the relevant publications, it took a very long time to arrive at this apparently trivial but conceptually revolutionary observation:

— 1545: Cardano publishes a solution to the cubic equation. Since some real solutions of this equation could not be expressed without the help of complex numbers, these numbers could no longer be rejected as 'impossible' — as they had been since the Babylonians.

— 1637: Descartes publishes his epochal "Discours sur la méthode", in which the geometric representation of pairs of numbers is introduced for the first time.

— 1799: Wentzel (a mathematical amateur) publishes the idea of the complex plane, but in Danish — a language in which his idea remains unnoticed for one hundred years.

— 1806: Argand (a mathematical amateur) self-publishes on the subject a pamphlet in French — which is at first appreciated by a few mathematicians, but soon forgotten.

— 1831: Gauss publishes in Latin his independent discovery of the complex plane — which is then universally accepted.

— 1833: Hamilton provides an abstract algebraic understanding of the complex numbers that frees them from the ontologically objectionable square root of minus one.

The following terminology is used interchangeably in the literature: Complex plane, Gaussian plane, and Argand plane. This plane is the direct product of the real and imaginary axes, whose characterization extends to algebras other than the field of complex numbers. A key concept is the operator $\mathcal{C}$ of complex (or Hermitian) conjugation. Since this operator is an involution, $\mathcal{C}^2 = I$, it has two eigenvalues, $+1$ and -1. The real an imaginary axes are the corresponding eigenspaces. This characterization is sufficiently general for immediate application to quantions:

Definition 47 *By **Gaussian space,** we shall refer to the direct product of the eigenspaces of the operator of complex conjugation.*

Independently of dimensionality, let us refer to such eigenspaces as the **real axis** and the **imaginary axis** and denote them by $\mathfrak{R}$ and $\mathfrak{J}$ respectively. For complex numbers, $\mathfrak{R} = \mathbb{R}$ and $\mathfrak{J} = i\mathbb{R}$. For quantions, $\mathfrak{R} = \mathcal{L}^h \left(= \mathbb{R}^4 \right)$ and $\mathfrak{J} = i\mathbb{R}^4$.

The Gaussian spaces for complex numbers, quaternions, and quantions, are comparatively illustrated in the following diagrams.

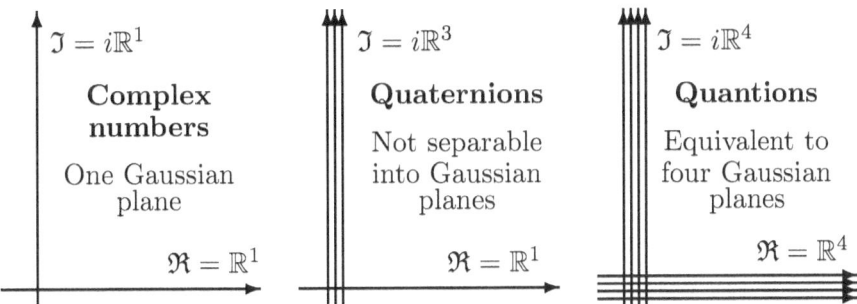

Figure 11.2: Gaussian spaces.

Every Gaussian space contains three metric structures. They are defined by the metric norms inside the real axis, inside the imaginary axis, and in the space itself.

For complex numbers, $z = x + iy$, all three norms are Euclidean. They are x^2, y^2, and $x^2 + y^2$ respectively.

For quaternions, $q = w + ix + jy + jz$, the three norms are also Euclidian. They are w^2 on the real axis, $x^2 + y^2 + z^2$ in the imaginary axis, and $w^2 + x^2 + y^2 + z^2$ in general.

For quantions, the metric structure of the Gaussian space is the algebraic source of relativity: The metric in $\mathfrak{R}$ (and consequently in $\mathfrak{J}$) is real Minkowskian. The metric in $\mathcal{M}$ is complex Minkowskian. These results are derived in the next section.

11.2 The relativistic metric

The quantionic Gaussian space is the four-dimensional linear space of complex vectors x^μ in the decomposition (2.18), that is,

$$P = x^\mu \Lambda_\mu = x^0 \Lambda_0 + x^1 \Lambda_1 + x^2 \Lambda_2 + x^3 \Lambda_3. \qquad (11.3)$$

Theorem 48 *The metric dual of $P = x^\mu \Lambda_\mu$ is*

$$P^\# = \sum_{\mu=0}^{3} x_\mu \Lambda_\mu = x^\mu \Lambda_\mu^\#. \qquad (11.4)$$

Proof. In matrix form,

$$P_{red}^\# = \begin{pmatrix} x^0 - x^3 & -x^1 + ix^2 \\ -x^1 - ix^2 & x^0 + x^3 \end{pmatrix}.$$

Hence

$$P^\# = x^0 \Lambda_0 - x^1 \Lambda_1 - x^2 \Lambda_2 - x^3 \Lambda_3 = \sum_{\mu=0}^{3} x_\mu \Lambda_\mu.$$

This proves the first part of relations (11.4).

Independently, we may also compute directly the duals of the lambda matrices:

1. Since $\Lambda_0 = I$, the dual is $\Lambda_0^\# = \Lambda_0$.
2. Since $(\Lambda_i)_{red} = \sigma_i$, the dual is

$$\Lambda_i^\# = -\Lambda_i. \qquad (11.5)$$

Thus, the metric dual coincides with the parity operator,

$$\Lambda_\mu^\# = \mathcal{P}\Lambda_\mu. \qquad (11.6)$$

This proves the second part of relations (11.4). ∎

Theorem 49 *For any two quantions $P, Q \in \mathcal{L}$, the inner product (P, Q) defines an inner product (x, y) in the linear space of coefficients by the relation*

$$(P, Q) = (x, y) I, \qquad (11.7)$$

where (x, y) is the Minkowskian scalar product

$$(x, y) \overset{def}{=} \eta_{\mu\nu} x^\mu y^\nu. \qquad (11.8)$$

Proof. Write

$$P \;=\; x^\mu \Lambda_\mu = x^0 I + \vec{x} \cdot \vec{\Lambda},$$
$$Q \;=\; y^\mu \Lambda_\mu = y^0 I + \vec{y} \cdot \vec{\Lambda}.$$

By Theorem 48, the expression for $P^\#$ assumes the simple form

$$P^\# = x^0 \Lambda_0^\# + x^1 \Lambda_1^\# + x^2 \Lambda_2^\# + x^3 \Lambda_3^\#, \tag{11.9}$$

or, equivalently

$$P^\# = \sum_{\nu=0}^{3} x_\nu \Lambda_\nu = x^0 - \vec{x} \cdot \vec{\Lambda},$$

and similarly for $Q^\#$:

$$Q^\# = \sum_{\nu=0}^{3} y_\nu \Lambda_\nu = y^0 - \vec{y} \cdot \vec{\Lambda}.$$

Hence, by relation (8.6),

$$
\begin{aligned}
(P,Q) \;&=\; \frac{1}{2}\left(P^\# Q + Q^\# P \right) \\
&=\; \frac{1}{2}\left(x^0 - \vec{x} \cdot \vec{\Lambda} \right)\left(y^0 + \vec{y} \cdot \vec{\Lambda} \right) + \frac{1}{2}\left(y^0 - \vec{y} \cdot \vec{\Lambda} \right)\left(x^0 + \vec{x} \cdot \vec{\Lambda} \right) \\
&=\; x^0 y^0 I - \frac{1}{2}\sum_{i,j=1}^{3} x_i y_j \left(\Lambda_i \Lambda_j + \Lambda_j \Lambda_i \right) \\
&=\; \left(x^0 y^0 - \sum_{i=1}^{3} x_i y_i \right) I \\
&=\; \left(x_\mu y^\mu \right) I \\
&\equiv\; (x,y)\, I,
\end{aligned}
$$

proving the theorem. ∎

Thus, the algebraic concept of a metric norm gives rise to a linear Minkowski space $M_0^4\,(\mathbb{C})$. This Gaussian space is linearly isomorphic with $\mathcal{L}$, but while the natural structure of $\mathcal{L}$ is that of an algebra, the natural structure of $M_0^4\,(\mathbb{C})$ is that of a metric space.

11.3   Inherent causality

The algebraic norm function $A(z) = z^*z$ introduces an asymmetry into the set of real numbers by assigning a distinguished role to the positive half of the real axis. This suggests a conceptual question: Should the norm function

$$A : \mathbb{C} \to \mathbb{R}^+, \tag{11.10}$$

that is,

$$A\left(re^{i\phi}\right) = r^2 \in \mathbb{R}^+, \tag{11.11}$$

be interpreted as a mapping from the field $\mathbb{C}$ into the field $\mathbb{R}$, or into the number system $\mathbb{R}^+$?

As already shown in Figure 1.1 on page 3, we adopt the second interpretation. Graphically:

$$
\begin{array}{ccccc}
\mathbb{R}^+ & \times & U(1) & = & \mathbb{C} \\
\Big\downarrow A & & \Big\downarrow A & & \Big\downarrow A \\
\mathbb{R}^+ & \times & 1 & = & \mathbb{R}^+
\end{array}
$$

By the theorems of Section 8.3 and Figure 8.1 on page 104, the algebraic norm function maps the algebra $\mathcal{L}$ of quantions onto the set $\mathcal{L}^+$ of Hermitian quantions with positive trace and non-negative algebraic norm. The norm function

$$A : \mathcal{L} \to \mathcal{L}^+ \tag{11.12}$$

is thus the quantionic extension of the complex norm function (11.10). Specifically, to the quantionic relation

$$A(ER) = R^2 \in \mathcal{L}^+, \tag{11.13}$$

corresponds the complex relation (11.11), where $E \in U_q(1)$ is an arbitrary phase factor and $R \in \mathcal{L}^+$ an arbitrary modulus.

Using the following mappings

$$
\left.
\begin{array}{c}
\mathbb{C} \rightleftarrows \mathcal{L}, \\
\mathbb{R}^+ \rightleftarrows \mathcal{L}^+, \\
U(1) \rightleftarrows U_q(1),
\end{array}
\right\}
\tag{11.14}
$$

where the arrows "$\leftarrow$" and "$\rightarrow$" indicate restriction and extension respectively, the graph of the algebraic norm function on page 132 generalizes to the commutative diagram

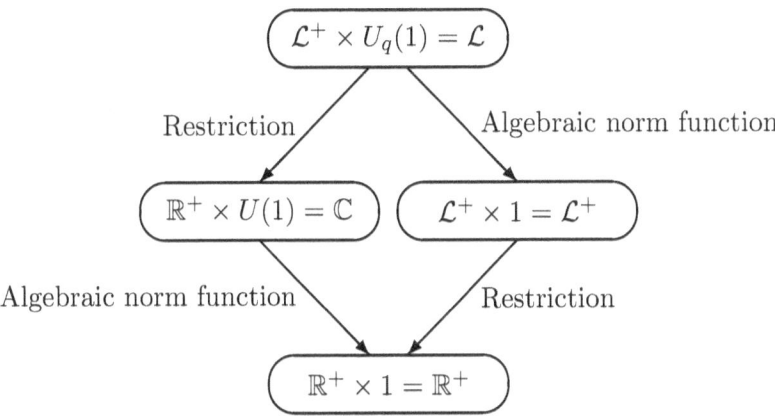

The algebraic norm function *introduces an asymmetry* into the algebra of quantions by assigning a distinguished meaning to the subset $\mathcal{L}^+$ of Hermitian quantions. We refer to this asymmetry as **inherent causality.**

The geometric interpretation of the concept of inherent causality is hierarchically illustrated by the five one-to-one mappings in Figure 11.3. Let us discuss each of these correspondences in turn.

The correspondence 1: This is the fundamental one-to-one correspondence (11.3) between the algebra $\mathcal{L}$ of quantions and a complex linear Minkowski space $M_0^4(\mathbb{C})$. We consider it 'fundamental' because it relates the algebra of quantions (quantum mechanics) to the geometric structure of spacetime (relativity). The mappings 2 to 5 are restrictions of this general mapping to various substructures of the algebra of quantions.

ALGEBRA   GEOMETRY

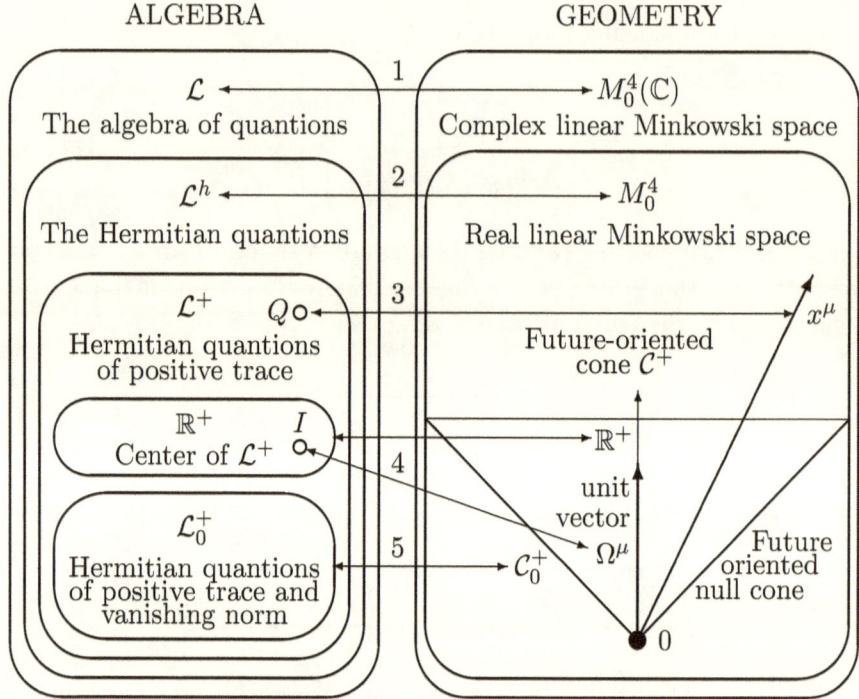

Figure 11.3. The algebraic source of causality.

The correspondence 2: Since the lambda matrices are Hermitian, it follows that a quantion $Q = x^\mu \Lambda_\mu$ is Hermitian, $Q^\dagger = Q$, if and only if the coefficients x^μ are real numbers, $(x^\mu)^* = x^\mu$. The linear subspace $\mathcal{L}^h \subset \mathcal{L}$ of Hermitian quantions is thus in one-to-one correspondence with a real linear Minkowski space $M_0^4 \subset M_0^4(\mathbb{C})$. As usual, M_0^4 is the shorter notation for $M_0^4(\mathbb{R})$.

The correspondence 3: The linear subspace $\mathcal{L}^+ \subset \mathcal{L}^h$ of Hermitian quantions Q with positive trace and positive metric norm,

$$\left. \begin{array}{r} Tr\,(Q) > 0, \\ \det\,(Q_{red}) > 0, \end{array} \right\} \tag{11.15}$$

is distinguished as the image of the algebra of quantions under the algebraic metric norm function,

$$A : \mathcal{L} \to \mathcal{L}^+. \tag{11.16}$$

On the other hand, the subspace $\mathcal{L}^+$ is in one-to-one correspondence with the set of time-like future-oriented Minkowski vectors $(x^\mu) \in C^+$.

In terms of the components x^μ, the relations (11.15) have a simple geometric interpretation:

$$\left.\begin{array}{ll} \text{Future-oriented vectors:} & Tr\,(Q) = 4x^0 > 0, \\[2mm] \text{Time-like vectors:} & \det\,(Q_{red}) = x_\mu x^\mu > 0. \end{array}\right\} \qquad (11.17)$$

Let us emphasize that the correspondence 3 is not the source of causality — even though this should be sufficiently clear from the foregoing. The reason is that there exists also a linear subspace $\mathcal{L}^- \in \mathcal{L}^h$ of Hermitian quantions with negative trace which corresponds to a past-oriented cone in M_0^4. But $\mathcal{L}^-$ is left out of Figure 11.3 because it is not structurally distinguished. In contrast, the subspace $\mathcal{L}^+$ is distinguished as the image of the full algebra $\mathcal{L}$ of quantions under the algebraic norm function. Thus, by relation (2.20), we have

$$\left.\begin{array}{l} A\,(Q) = j^\mu \Lambda_\mu, \\[2mm] j^\mu = \frac{1}{4} Tr(A\,(Q)\,\Lambda_\mu). \end{array}\right\} \qquad (11.18)$$

The correspondence 4: The basis quantion Λ_0 is distinguished in the algebra $\mathcal{L}$ as the unit element, $\Lambda_0 \equiv I$, and, since $I \in \mathcal{L}^h$, it follows that the four-vector which corresponds to I is distinguished in the Minkowski space M_0^4. We shall denote this vector by Ω^μ, and refer to it as **the structure vector.** The linear decomposition

$$\Lambda_0 = \Omega^\mu \Lambda_\mu$$

implies $\Omega^0 = 1$ and $\vec{\Omega} = 0$, that is,

$$(\Omega^\mu) = \begin{pmatrix} 1 \\ 0 \\ 0 \\ 0 \end{pmatrix}. \qquad (11.19)$$

Thus, the structure vector Ω^μ defines the future-oriented time-like direction in the linear Minkowski space M_0^4. The positive half-axis $\mathbb{R}^+$ is the geometric object which corresponds to the center of $\mathcal{L}$, that is, to the field of complex numbers, $\mathbb{C}I \subset \mathcal{L}$. As an algebraic object, $\mathbb{R}^+$ is the center of $\mathcal{L}^+$.

The correspondence 5: To the set $\mathcal{L}_0^+$ of Hermitian quantions Q with positive trace and vanishing metric norm,

$$
\begin{aligned}
Tr\,(Q) &= 4x^0 > 0, \\
\det\,(Q_{red}) &= x_\mu x^\mu = 0,
\end{aligned}
$$

corresponds the null cone $\mathcal{C}_0^+$ of future-oriented null vectors in the linear Minkowski space M_0^4.

This completes the discussion of Figure 11.3.

Returning to the ideal $\mathcal{J}$ defined in Section 8.3, we note that the subset $\mathcal{L}_0^+$ of quantions is the real part of $\mathcal{J}$. We thus have the relations

$$
\mathcal{C}_0^+ \longleftrightarrow \mathcal{L}_0^+ \subset \mathcal{J} \subset \mathcal{L}.
$$

In general, the ideal which generalizes the element zero of any algebra 'measures' by how much this algebra fails to be a division algebra. It follows that an algebra is a division algebra if and only if the ideal which generalizes the element zero contains no element other than zero.

The light cone $\mathcal{C}_0^+$ is thus the geometric counterpart (or physical interpretation) of the ideal $\mathcal{J}$ by which the algebra of quantions differs from the field of complex numbers from the viewpoint of the division structure defined on page 15 (see also Table 1.1 on page 14).

A discussion of unification

Understood as a research program, the phrase "unification of relativity and quantum mechanics" brings to mind a hypothetical construction similar to Maxwell's unification of electrical and magnetic phenomena, or to Einstein's attempted unification of the gravitational and electromagnetic fields. This view of unification is illustrated in Figure 11.4: Given two physical theories, the object is to find a new theory that contains the given theories as substructures.

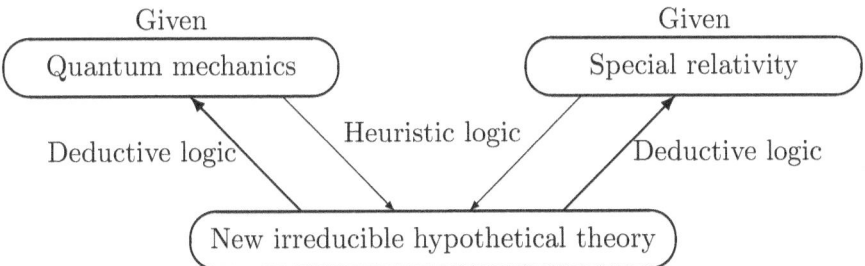

Figure 11.4. Unification as a research program.

The essence of this research paradigm is that the new theory has to be guessed within certain conditions. We refer to the arguments that support the guess as "heuristic logic". Unlike deductive logic, which yields *exact theorems,* heuristic logic yields *plausible conclusions* that must be validated by deductive logic.

As it is only the end result that matters, a unification need not be the outcome of the program outlined in Figure 11.4. Figure 11.5. illustrates the quantionic approach.

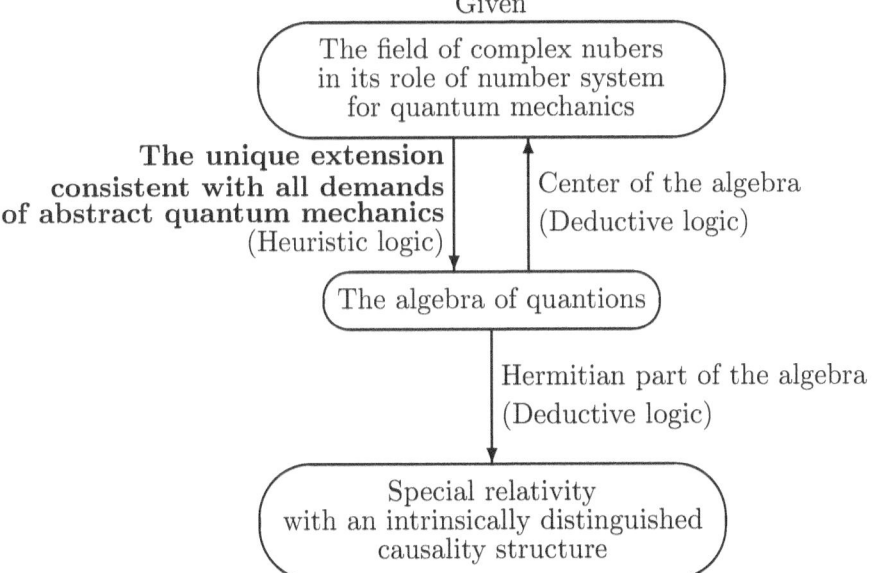

Figure 11.5. Unification as a heuristic theorem.

In this approach, the transition from the field of complex numbers to the algebra of quantions is also based on heuristic arguments, but, as is evident from Table 1.1 on page 14, these arguments are so strong that they apparently leave no room for choice.

The algebra of quantions is thus the extension of the field of complex numbers which satisfies all conditions imposed by the abstract structure of quantum mechanics. This extension is unique.

11.4   The tensor formalism

The present section is included only to help clarify the meaning of the structure vector Ω^μ in Figure 11.3. It may be ignored because the tensor formalism is not needed in the remainder of this book.

****

Since the matrix components q_i will not be needed, the vectors x^μ may be written as q^μ without confusion.

In covariant form, the definition (11.19) reads

$$\left.\begin{array}{c} (\Omega, \Omega) = 1, \\ (\omega|\Omega|\omega) = \frac{1}{2}. \end{array}\right\} \tag{11.20}$$

The following theorem is at the basis of the tensorial formalism.

Theorem 50 *For two arbitrary quantions, $P = p^\mu \Lambda_\mu$ and $Q = q^\mu \Lambda_\mu$, the vector w^μ which corresponds to the product PQ, that is, $PQ = w^\mu \Lambda_\mu$, is given by the relation*

$$\begin{aligned} w^\mu &= (\Omega, q)\, p^\mu + (\Omega, p)\, q^\mu - (p, q)\, \Omega^\mu \\ &\quad - i\eta^{\mu\alpha} \varepsilon_{\alpha\beta\gamma\delta} \Omega^\beta p^\gamma q^\delta. \end{aligned} \tag{11.21}$$

Proof. We are to compute the product

$$PQ = (p^\mu \Lambda_\mu)(q^\nu \Lambda_\nu) = p^\mu q^\nu \Lambda_\mu \Lambda_\nu$$

by expanding the object $\Lambda_\mu \Lambda_\nu$. Written out in full, the product reads:

$$\begin{aligned} &PQ \\ &= \left(p^0 \Lambda_0 + p^1 \Lambda_1 + p^2 \Lambda_2 + p^3 \Lambda_3\right)\left(q^0 \Lambda_0 + q^1 \Lambda_1 + q^2 \Lambda_2 + q^3 \Lambda_3\right) \\ &= p^0 q^0 \Lambda_0 + p^0 \left(q^i \Lambda_i\right) + q^0 \left(p^i \Lambda_i\right) + \left(p^1 q^1 + p^2 q^2 + p^3 q^3\right) \Lambda_0 \\ &\quad + i\left(p^2 q^3 - p^3 q^2\right) \Lambda_1 + i\left(p^3 q^1 - p^1 q^3\right) \Lambda_2 + i\left(p^1 q^2 - p^2 q^1\right) \Lambda_3. \end{aligned}$$

The first two terms of the third line equal $p^0 q^\mu \Lambda_\mu$.

By adding $q^0 p^0 \Lambda_0$, the third term of the third line becomes $q^0 p^\nu \Lambda_\nu$.

By removing $q^0 p^0 \Lambda_0$, the fourth term of the third line becomes $- (q^\mu p_\mu) \Lambda_0$. Furthermore, the unit matrix Λ_0 may be written as $\Omega^\mu \Lambda_\mu$. Hence, the third line simplifies to

$$\text{line } 3 = p^0 \left(q^\mu \Lambda_\mu \right) + q^0 \left(p^\mu \Lambda_\mu \right) - \left(p_\nu q^\nu \right) \Omega^\mu \Lambda_\mu.$$

Using the Levi-Civita symbol, the fourth line may be condensed into the form

$$\text{line } 4 = i \delta^{kl} \varepsilon_{ijk} q^i p^j \Lambda_l.$$

In the four-dimensional formalism, characterized by Greek indices, we have $(\varepsilon_{ijk}) = (\varepsilon_{0\lambda\mu\nu})$. Hence:

$$\text{line } 4 = -i \eta^{\nu\mu} \varepsilon_{0\kappa\lambda\mu} p^\kappa q^\lambda \Lambda_\nu.$$

The minus sign comes from the metric tensor: $\eta^{kl} = \eta_{kl} = -\delta_{kl}$.

Collecting these results, we have:

$$PQ = \left(p^0 q^\mu + q^0 p^\mu - q^\nu p_\nu \Omega^\mu - i\eta^{\nu\mu} \varepsilon_{0\kappa\lambda\mu} p^\kappa q^\lambda \right) \Lambda_\nu.$$

The term in parentheses is the vector w^μ, but we still have to eliminate the index 0. This can be done by the following observation: every occurrence of the index 0 can be replaced by a dummy index tied to the index of the structure vector Ω. Thus,

$$\begin{aligned} q^0 &= q^\nu \Omega_\nu = q_\nu \Omega^\nu, \\ \varepsilon_{0\kappa\lambda\nu} q^\kappa p^\lambda &= \varepsilon_{\alpha\kappa\lambda\nu} p^\kappa q^\lambda \Omega^\alpha. \end{aligned}$$

After some aesthetic rearrangements, one obtains relation (11.21). ∎

Note: The proof of relation (11.21) is much shorter in terms of 3-vectors:

$$\begin{aligned} PQ &= \left[p^0 \Lambda_0 - \vec{p} \cdot \vec{\Lambda} \right] \left[q^0 \Lambda_0 - \vec{q} \cdot \vec{\Lambda} \right] \\ &= \left(p^0 q^0 + \vec{p} \cdot \vec{q} \right) \Lambda_0 \\ &\quad + \left(-q^0 \vec{p} - p^0 \vec{q} + i\vec{p} \times \vec{q} \right) \cdot \vec{\Lambda}. \end{aligned}$$

By its construction, the right-hand side of relation (11.21) is manifestly covariant, but it contains the structure vector Ω^μ. Due to the presence of such a distinguished vector (a field of vectors in general relativity) some physicists have objected to the algebra of quantions for not being manifestly covariant. The author's opinion is that manifest covariance and the presence of a distinguished tensorial object (the structure vector Ω^μ) are unrelated issues. It is only if the distinguished vector field is used to specify a coordinate system that manifest covariance is lost. Thus, relation (11.21) is manifestly covariant, but its second proof in terms of 3-vectors is not because Ω^μ has been identified with the direction of the time axis.

As for the presence of the distinguished vector Ω^μ, it is true that it is an object which does not appear in general relativity — but it is also true that general relativity does not include quantum phenomena. Our contention is that quantum phenomena require the existence of a distinguished time-like future oriented direction at every point of spacetime.

To verify the integrity of relation (11.21), let us prove causality in the tensorial formalism.

Substitution of $p^{*\mu}$ for q^μ in relation (11.21) yields the vector j^μ which corresponds to the norm $A(Q)$:

$$\begin{aligned} j^\mu &= (\Omega, p^*)\, p^\mu + (\Omega, p)\, p^{*\mu} - (p, p^*)\, \Omega^\mu \\ &\quad - i\eta^{\mu\alpha}\varepsilon_{\alpha\beta\gamma\delta}\Omega^\beta p^\gamma p^{*\delta}. \end{aligned} \tag{11.22}$$

To verify that this vector is real, $j^{*\mu} = j^\mu$, we consider the effects of complex conjugation on the right-hand side:

(a) conjugation interchanges the first two terms,

(b) conjugation leaves the third term invariant,

(c) conjugation changes of sign of the imaginary unit in the fourth term, but this is compensated by the transposition $\gamma \rightleftarrows \delta$ of indices in the antisymmetric Levi-Civita symbol.

To compute the Minkowski norm (j, j), we write j^μ as a sum of two vectors:

$$j^\mu = F^\mu + G^\mu,$$

where

$$\begin{aligned} F^\mu &= (\Omega, p^*)\, p^\mu + (\Omega, p)\, p^{*\mu} - (p, p^*)\, \Omega^\mu, \\ G^\mu &= -i\eta^{\mu\alpha}\varepsilon_{\alpha\beta\gamma\delta}\Omega^\beta p^\gamma p^{*\delta}. \end{aligned}$$

The motivation for doing so is that $(F, G) = 0$. Indeed, the three terms of this scalar product are proportional, respectively, to the the three expressions

$$\varepsilon_{\alpha\beta\gamma\delta}\Omega^\beta p^\gamma p^{*\delta} p^\alpha,$$
$$\varepsilon_{\alpha\beta\gamma\delta}\Omega^\beta p^\gamma p^{*\delta} p^{*\alpha},$$
$$\varepsilon_{\alpha\beta\gamma\delta}\Omega^\beta p^\gamma p^{*\delta} \Omega^\alpha,$$

but all vanish identically due to the antisymmetry of $\varepsilon_{\alpha\beta\gamma\delta}$. Hence:

$$(j, j) = (F, F) + (G, G).$$

For the first term, direct computation yields

$$
\begin{aligned}
(F, F) \quad = \quad & (\Omega, p^*)^2 (p, p) + (\Omega, p)^2 (p, p)^* + (p, p^*)^2 \\
& -2 (\Omega, p^*) (\Omega, p) (p, p^*).
\end{aligned}
$$

To expand the second term,

$$(G, G) = -\eta^{\mu\alpha} \varepsilon_{\alpha\beta\gamma\delta} \Omega^\beta p^\gamma p^{*\delta} \varepsilon_{\mu\rho\sigma\tau} \Omega^\rho p^\sigma p^{*\tau},$$

we use the identity[1]

$$
\begin{aligned}
-\eta^{\mu\alpha} \varepsilon_{\alpha\beta\gamma\delta} \varepsilon_{\mu\rho\sigma\tau} \quad = \quad & \eta_{\beta\rho}\eta_{\gamma\sigma}\eta_{\delta\tau} + \eta_{\beta\tau}\eta_{\gamma\rho}\eta_{\delta\sigma} + \eta_{\beta\sigma}\eta_{\gamma\tau}\eta_{\delta\rho} \\
& -\eta_{\beta\rho}\eta_{\gamma\tau}\eta_{\delta\sigma} - \eta_{\beta\sigma}\eta_{\gamma\rho}\eta_{\delta\tau} - \eta_{\beta\tau}\eta_{\gamma\sigma}\eta_{\delta\rho}.
\end{aligned}
$$

Its substitution into the expression for (G, G) yields

$$
\begin{aligned}
(G, G) \quad = \quad & (p, p) (p, p)^* + 2 (\Omega, p^*) (\Omega, p) (p, p^*) \\
& - (p, p^*)^2 - (\Omega, p)^2 (p, p)^* - (\Omega, p^*)^2 (p, p).
\end{aligned}
$$

In the sum of (F, F) and (G, G), all terms cancel except one:

$$(j, j) = (p, p) (p, p)^*. \tag{11.23}$$

If the Minkowski norm (p, p) is either positive or negative, the norm (j, j) is positive. It vanishes only if $(p, p) = 0$. In other words, the vector j may be time-like or light-like, but not space-like.

---

[1]Up to an arbitrary coefficient, the right-hand side is the most general antisymmetric expression for $\varepsilon_{\mu\rho\sigma\tau}$. The coefficient is uniquely defined by the identity $\varepsilon_{\alpha\beta\gamma\delta}\varepsilon^{\alpha\beta\gamma\delta} = -4!$

To verify that j^μ is future-oriented, one shows that its projection (Ω, j) on the time axis Ω^μ is positive. Relation (11.22) yields

$$
\begin{aligned}
(\Omega, j) &= (\Omega, p^*)(\Omega, p) + (\Omega, p)(\Omega, p^*) - (p, p^*) \\
&= 2(\Omega, p)^*(\Omega, p) - (p, p^*).
\end{aligned}
$$

Let us write p^μ in the form

$$
p^\mu = z\Omega^\mu + w^\mu,
$$

where z is a complex number and $w^\mu = p^\mu + iv^\mu$, the two real vectors p^μ and q^μ being orthogonal to Ω^μ,

$$
(\Omega, w) = 0.
$$

Hence:

$$
(w, w^*) = (p, p) + (q, q) \leqslant 0.
$$

This means that w^μ is space-like. Since

$$
\begin{aligned}
(\Omega, p) &= z, \\
(p, p^*) &= z^* z + (w, w^*),
\end{aligned}
$$

one obtains

$$
(\Omega, j) = 2z^* z - z^* z - (w, w^*) = z^* z - (w, w^*) > 0, \tag{11.24}
$$

which completes the tensorial proof of causality.

Chapter 12

Quantionic Hilbert Spaces

The present chapter may be ignored in the first reading because its only purpose is to support the statement, made on several occasions, that a generalization of Hilbert space can be built over the algebra of quantions without conceptually modifying any basic ideas. It is by this property that the algebra of quantions enables us to break through the rigidity of the standard Hilbert space — a 'break-through' needed to make quantum mechanics relativistic. But since Part II is limited to the derivation of the quantionic equations of motion for single-component fields, quantionic Hilbert spaces of more than one quantionic dimension play no role in the present work.

****

A "quantionic Hilbert space" is the mainimal structure which becomes a standard complex Hilbert space if one restricts the algebra of quantions to its center.

Since the algebra of quantions is structurally richer than the field of complex numbers, the objects in quantionic Hilbert spaces enjoy some properties that have no counterpart in the standard Hilbert space.

We shall discuss only those concepts that are self-evident generalizations of standard ones. It is also sufficient to consider spaces of a finite number of dimensions.

Let $\mathcal{H}^n$ denote an n-dimensional quantionic Hilbert space. Its definition, as well as the definition of the unitary and Hermitian operators that act in this space, are immediate generalizations of the definitions of the corresponding standard objects: It suffices to substitute quantions for complex numbers and the Hermitian conjugation of quantions for the complex conjugation of complex numbers.

A quantionic ket, denoted by $|*\rangle$, is thus of the form

$$|Q\rangle = \begin{pmatrix} Q_1 \\ Q_2 \\ \vdots \\ Q_n \end{pmatrix} \in \mathcal{H}^n, \tag{12.1}$$

the components being arbitrary quantions, $Q_i \in \mathcal{L}$. The corresponding bra is

$$\langle Q| = \begin{pmatrix} Q_1^\dagger & Q_2^\dagger & \cdots & Q_n^\dagger \end{pmatrix}. \tag{12.2}$$

The norm of quantionic vectors

The norm

$$\langle \psi|\psi \rangle = \sum_{i=1}^{n} A(z_i) = \sum_{i=1}^{n} z_i^* z_i$$

in the complex Hilbert space generalizes directly to

$$\langle Q|Q \rangle = \sum_{i=1}^{n} A(Q_i) = \sum_{i=1}^{n} Q_i^\dagger Q_i \tag{12.3}$$

in the quantionic space $\mathcal{H}^n$.

Since a sum of positive real numbers, $A(z_i)$, is a positive real number, the norm $\langle \psi|\psi \rangle$ is positive-definite: $\langle \psi|\psi \rangle \in \mathbb{R}^+$. The analogous relation in the quantionic Hilbert space is

$$\langle Q|Q \rangle \in \mathcal{L}^+ \longleftrightarrow \mathcal{C}^+,$$

where $\mathcal{C}^+$ is the cone of future oriented vectors (time-like and light-like). This is true because the cone $\mathcal{C}^+$ is stable under addition and scaling by positive factors. The proof is straightforward: Take any two future-oriented Minkowski vectors $u, v \in \mathcal{C}^+$ and any two positive real numbers $r, s \in \mathbb{R}^+$. The norm of the vector

$$w = ru + sv,$$

which is defined only by positive scaling and addition, is

$$(w, w) = (ru + sv, ru + sv)$$
$$= r^2 (u, u) + s^2 (v, v) + 2rs (u, v).$$

Since all three terms on the right-hand side are positive definite, w is a time-like vector. And since $w^0 = ru^0 + sv^0 > 0$, the vector w is future-oriented. Thus, $w \in C^+$, which may be stated in a suggestive compact form:

The inherent causality of the algebraic norm of quantions propagates to the norm of quantionic Hilbert space vectors.

The quantionic unitary group

The most general quantionic linear transformation in $\mathcal{H}^n$,

$$|Q\rangle \mapsto |Q'\rangle = U |Q\rangle,$$

reads

$$Q_i \mapsto Q_i' = \sum_{j=1}^{n} U_{ij} Q_j, \tag{12.4}$$

where the n^2 coefficients $U_{ij} \in \mathcal{L}$ are arbitrary quantions.

The components of the corresponding bra are

$$Q_i'^{\dagger} = \sum_{k=1}^{n} (U_{ik} Q_k)^{\dagger} = \sum_{k=1}^{n} Q_k^{\dagger} U_{ik}^{\dagger}.$$

The transformed inner product is thus

$$\langle Q' | Q' \rangle = \sum_{i,j,k=1}^{n} Q_k^{\dagger} U_{ik}^{\dagger} U_{ij} Q_j \equiv \sum_{j,k=1}^{n} Q_k^{\dagger} \left(\sum_{i=1}^{n} U_{ik}^{\dagger} U_{ij} \right) Q_j.$$

If one now requires that the inner product be invariant,

$$\langle Q | Q \rangle' \equiv \langle Q | Q \rangle,$$

this identity implies

$$\sum_{i=1}^{n} U_{ik}^{\dagger} U_{ij} = \delta_{ij}. \tag{12.5}$$

In this expression, the dagger on the coefficient $U_{ik}^\dagger$ can mean only one thing: Hermitian conjugation in the algebra of quantions. By combining this conjugation with the transposition of the matrix U, we may write

$$U_{ik}^\dagger = (U_{ki})^\dagger,$$

where the dagger on the right-hand side represents Hermitian conjugation of the matrix and of its elements. Relation (12.5) now reads

$$\sum_{i=1}^{n} (U_{ki})^\dagger U_{ij} = \delta_{ij},$$

or, symbolically, $U^\dagger U = I$. Multiplying both sides of this relation from the left by U yields

$$UU^\dagger U = U,$$

which implies

$$UU^\dagger = I + Z,$$

where Z is some quantion such that

$$ZU = 0.$$

Since no unitary matrix is singular, meaning that U^{-1} always exists, multiplication of this relation from the right by U^{-1} yields $Z = 0$, and hence $UU^\dagger = I$.

A quantionic unitary matrix is thus formally defined by the same identities as a complex unitary matrix:

$$U^\dagger U \equiv UU^\dagger \equiv I. \tag{12.6}$$

The group properties are self-evident:

(1) The identity transformation is a unitary matrix.

(2) The inverse exists: $U^{-1} = U^\dagger$.

(3) Stability is guaranteed:

$$(UV)^\dagger (UV) = V^\dagger U^\dagger UV = V^\dagger IV = I.$$

(4) Associativity is a consequence of the associativity of the product of matrices and of the product of quantions.

Commutativity is doubly precluded because neither the product of quantions nor the product of matrices is commutative.

We shall refer to the group $U_q(n)$ of quantionic unitary transformations in n dimensions as the **quantionic unitary group.**

The quantionic Hermitian matrices

Let us consider an infinitesimal unitary matrix,

$$U_\varepsilon = I + \varepsilon A$$

for some quantionic matrix A. The unitarity condition (12.6) implies

$$U_\varepsilon^\dagger U_\varepsilon = (I + \varepsilon A)^\dagger (I + \varepsilon A) = I + \varepsilon \left(A^\dagger + A\right) + \varepsilon^2 A^\dagger A = I.$$

which further implies $A^\dagger + A = 0$. This means that, to first order, A is an arbitrary antihermitian matrix whose components are quantions.

The most general infinitesimal unitary transformation is thus

$$U_\varepsilon = I + \varepsilon i H, \tag{12.7}$$

where H is Hermitian,

$$H^\dagger = H. \tag{12.8}$$

Integration yields

$$U = \exp(iH) = \sum_{k=0}^{\infty} \frac{1}{k!} (iH)^n. \tag{12.9}$$

We see that there is no formal difference between the quantionic and complex cases in the definition of Hermitian matrices and in their relation to unitary matrices.

Singular vectors

As shown in Section 11.3, the algebraic norm of a singular quantion is a future-oriented Minkowski null vector. By analogy, we shall say that a Hilbert space vector $|Q\rangle \in \mathcal{H}^n$ is singular if its norm $\langle Q|Q\rangle$ is a Minkowski null vector.

Since the sum of future-oriented null vectors is a null vector only if all the summands are collinear, it follows that

$$\langle Q|Q\rangle = A(Q_1) + A(Q_2) + \cdots + A(Q_n), \tag{12.10}$$

is a null vector if and only if all Q_i are proportional to some singular vector Q_0,

$$Q_i = z_i Q_0,$$

where $(Q_0, Q_0) = 0$ (the singularity condition), and the coefficients z_i are arbitrary complex numbers. The expression (12.10) now reads

$$\langle Q|Q\rangle = (z_1^* z_1 + z_2^* z_2 + \cdots + z_n^* z_n)\, A\,(Q_0).$$

Since the real coefficient $(z_1^* z_1 + \cdots + z_n^* z_n)$ may be absorbed by Q_0 as a square root, we may write

$$\langle Q|Q\rangle = A\,(Q_0).$$

This result yields the following two insights:

(1) A singular vector in $\mathcal{H}^n$ is defined by two objects: a unit vector

$$|\psi\rangle = \begin{pmatrix} z_1 \\ z_2 \\ \vdots \\ z_n \end{pmatrix}$$

in a standard complex Hilbert space, and an arbitrary singular quantion Q_0. Let's refer to Q_0 as the "characteristic null quantion".

(2) The sum of two vectors in $\mathcal{H}^n$ is a singular vector if and only if the summands are singular and belong to the same characteristic null quantion.

An observation concerning eigenvalues

Since the eigenvalues of complex Hermitian matrices are real numbers, the eigenvalues of quantionic Hermitian matrices ought to be Hermitian quantions, that is, real Minkowski vectors. They are not in general. It follows that the classes of quantionic Hermitian matrices for which this is true may be expected to have interesting physical interpretations.

Chapter 13

Analysis

According to Robinson's intuitive characterization of number systems on page 2, the algebra of quantions may be regarded as a number system from the point of view of mathematics. This is evident from the graph on page 3 and from Table 1.1 on page 14.

According to the criterion suggested on top of page 7, the algebra of quantions may also be regarded as a number system from the point of view of physics if quantum-relativistic states can be represented by quantionic fields. Since this is a question in physics, not in mathematics, it belongs to Part II. As we shall see, the answer is positive.

We now observe that derivation operators exist in the standard number systems of physics ($\frac{d}{dx}$ in $\mathbb{R}$ and $\frac{d}{dz}$ in $\mathbb{C}$), but not in the other division algebras (this is shown on page 164 for quaternions). In the present chapter, we shall prove that an intrinsically distinguished derivation operator does exist in the algebra of quantions. Hence, derivation operators exist in the three number systems of physics and nowhere else — which means, suggestively: only where they are needed to formulate the differential equations of motion for physical states.

Owing to the structural richness of the algebra of quantions, the quantionic derivation operator, which will be denoted by $\mathcal{D}$, differs markedly from the familiar operator $\frac{d}{dz}$, so that new intuitions will have to be developed for quantionic analysis.

The quantionic derivation operator will be constructed in the three steps represented by arrows in Figure 13.1. The three sections of this chapter are dedicated in turn to each of these steps.

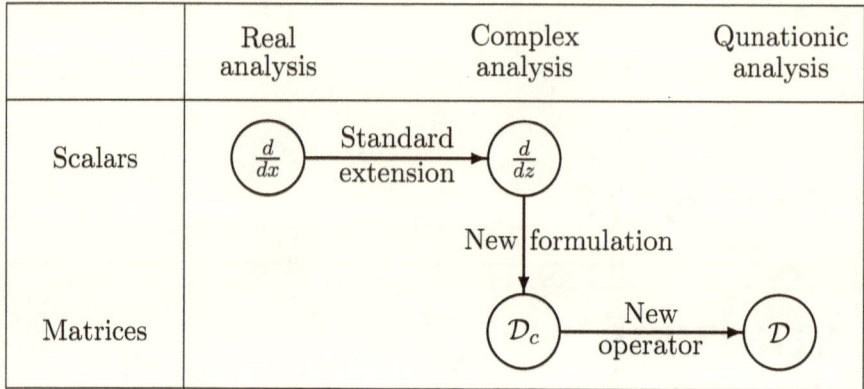

Figure 13.1. The path to quantionic derivatives.

The objective is to construct a quantionic derivation operator $\mathcal{D}$ which generalizes the real derivation operator $\frac{d}{dx}$. Except for linearity, the properties of the hypothetical operator $\mathcal{D}$ are initially unknown.

The construction strategy is based on two observations:

1. The derivation operator in the field of real numbers is not rich enough structurally to suggest a sufficient number of defining properties for a derivation operator in the algebra of quantions. But since the field of complex numbers is a substructure of the algebra of quantions, complex analysis is a natural intermediate step for the generalization of real analysis to quantionic analysis.

2. Both real and complex analysis are 'scalar theories', in the sense that the variables are the real scalar x and the complex scalar z. On the other hand, quantions are 4×4 matrices. We shall see that the transition from scalars to matrices can be effected first in complex analysis, and then extended to quantions.

These observations are developed in the following three steps:

Section 13.1: The extension of real analysis to the complex analysis of analytic functions is reviewed in this section in a somewhat unusual approach that brings to light the essential properties of the derivative. Five such properties are identified.

Section 13.2: The arrow labeled "New formulation" represents a reformulation of complex analysis in the formalism of 2×2 matrices, where $\frac{d}{dz}$ is represented by a matrix operator denoted by $\mathcal{D}_c$.

Section 13.3: The arrow "New operator" represents the procedure that extends the complex operator $\mathcal{D}_c$ to a quantionic operator $\mathcal{D}$, which is a 4×4 matrix. This procedure yields the concrete expression for the quantionic derivation operator $\mathcal{D}$.

13.1   Derivation in the field of complex numbers

Analysis was initially developed for real functions, where a tangent is associated to every point of a smooth curve by the definition

$$\frac{d}{dx} f\left(x\right) = \lim_{h \to 0} \frac{f\left(x+h\right) - f\left(x\right)}{h} \tag{13.1}$$

of the derivation operator $\frac{d}{dx}$. The domain and range of the function f are open intervals in the field $\mathbb{R}$ of real numbers, or simply

$$f : \mathbb{R} \to \mathbb{R}. \tag{13.2}$$

In differential geometry and quantum field theory, one considers simultaneously more than one independent variable and more than one dependent variable:

$$f : \mathbb{R}^n \to \mathbb{R}^m. \tag{13.3}$$

In the presence of symmetry groups, the partial derivatives are generalized to covariant derivatives, but these are straightforward compensations for the modifications implied by the groups in question. In contrast, the transition to a new number system modifies the concept of derivation in a more profound way. Thus, the extension of the mapping (13.2) to complex numbers,

$$w : \mathbb{C} \to \mathbb{C}, \tag{13.4}$$

calls for a new approach to derivatives. This is why nearly two centuries elapsed between the discoveries of real and complex analysis.

If we ignore the algebraic structure of the complex numbers, the function (13.4) reduces to a mapping of type (13.3), that is, to

$$f : \mathbb{R}^2 \to \mathbb{R}^2. \tag{13.5}$$

This is just a set of two real functions of two real variables. It is conventionally written in the form

$$
\left.
\begin{array}{l}
u = u\left(x, y\right), \\
v = v\left(x, y\right).
\end{array}
\right\}
\tag{13.6}
$$

In order to view the field $\mathbb{C}$ of complex numbers as a number system, the function (13.4) is to be treated as the assignment $z \mapsto w$ of *one number* to *one number* — not as the assignment $(x, y) \mapsto (u, v)$ of a *pair of numbers* to a *pair of numbers*. The function (13.4) may nevertheless be written in the form (13.6), provided that the Cauchy-Riemann equations are satisfied.

The textbook approach to complex analysis begins by extending the definition (13.1) to complex functions.[1] This is possible because the properties of the field of real numbers needed for (13.1) to be meaningful are also present in the field of complex numbers. They are: associativity and commutativity (they guarantee a unique expansion of $w\left(z + h\right)$), and division structure (it guarantees the existence of the ratio). Hence, the derivative of a complex function $w\left(z\right)$ is

$$
\frac{d}{dz} w\left(z\right) \stackrel{def}{=} \lim_{h \to 0} \frac{w\left(z + h\right) - w\left(z\right)}{h},
\tag{13.7}
$$

where h is an auxiliary complex variable.

The above emphasis on necessary conditions is to point out that the definition (13.1) does not extend to the algebra of quantions for lack of commutativity and division structure.

The transition from the derivative (13.1) of real functions to the derivative (13.7) of complex functions exhibits two novelties: the loss of geometric interpretation and the loss of uniqueness.

Loss of interpretation: The visualization of the derivative as a tangent played an essential role in the development of the differential calculus. It is also essential in its physical applications. Since this interpretation is absent in algebras of more than one real dimension,

---

[1] Very thorough classical texts based on the algorithmic approach developed in the 19th century are: Whittaker & Watson, *A Course of Modern Analysis.* Cambridge University Press, Cambridge.(many editions since 1902), and Forsyth, *Theory of Functions of a Complex Variable,* in two volumes, Cambridge University Press, Cambridge (1918).

more abstract criteria must be found as a foundation for complex analysis.

Loss of uniqueness: The limit (13.1) is well defined for all real smooth functions, but the limit (13.7) is unique only for a very special class of complex functions. To see why, we write

$$h = re^{i\alpha}.$$

The limit $h \to 0$ is then equivalent to $r \to 0$, so that (13.7) becomes

$$\frac{d}{dz}w(z) = \lim_{r \to 0} \frac{w(z + re^{i\alpha}) - w(z)}{re^{i\alpha}}. \tag{13.8}$$

While the phase factor $e^{i\alpha}$ plays no role in taking the limit, the function $w'(z) = \frac{d}{dz}w(z)$ may nevertheless be a function of α. If this happens to be the case, this function is not a function of the single complex variable z, but of the pair of variable $\{z, \alpha\}$. This means that the field $\mathbb{C}$ is not viewed as a single number system, but as a two-dimensional linear space $\mathbb{R}^2$. From the viewpoint of analysis, the number system remains $\mathbb{R}$.

For the field $\mathbb{C}$ to be a number system from the point of view of analysis, the set of admissible functions

$$w(z) = u(x, y) + iv(x, y) \tag{13.9}$$

must be limited to those functions whose derivatives are not direction-dependent, which means that the phase factor $e^{i\alpha}$ cancels out in the limit. For such functions, the derivative $\frac{d}{dz}w(z)$ is uniquely defined.

Definition 51 *A complex function $w(z)$ is **analytic** at a point z if and only if its derivative (13.8) is not a function of α :*

$$\frac{\partial}{\partial \alpha}\left(\frac{d}{dz}w(z)\right) \equiv 0. \tag{13.10}$$

The following theorem is fundamental in the theory of complex analytic functions:[2]

---

[2]Other terms, synonymous to "analytic" in the absence of singularities, are **holomorphic** and **monogenic.** The term "analytic" is chosen in the present work for being the most commonly used in physics texts.

Theorem 52 *If the functions $w_1(z)$ and $w_2(z)$ are analytic and not singular a point z, their linear combinations*

$$w(z) = \lambda w_1(z) + \mu w_2(z) \tag{13.11}$$

and their product

$$w(z) = w_1(z) w_2(z) \tag{13.12}$$

are also analytic and not singular.

Proof. Relation (13.11) merely expresses the fact that the derivation operator $\mathcal{D}$ is linear.

To prove relation (13.12), we simplify the expression

$$\frac{d}{dz}(w_1 w_2) = \lim_{r \to 0} \frac{w_1(z + re^{i\alpha})\, w_2(z + re^{i\alpha}) - w_1(z)\, w_2(z)}{re^{i\alpha}}.$$

From relation (13.8) follows

$$\lim_{r \to 0} \left\{ w(z + re^{i\alpha}) = w(z) + re^{i\alpha} \frac{d}{dz} w(z) \right\}. \tag{13.13}$$

Substitution of this relation into the previous equation yields (dropping the arguments of the functions w_1 and w_2):

$$\frac{d}{dz}(w_1 w_2) = \lim_{r \to 0} \frac{\left(w_1 + re^{i\alpha} \frac{d}{dz} w_1\right)\left(w_2 + re^{i\alpha} \frac{d}{dz} w_2\right) - w_1 w_2}{re^{i\alpha}}$$

$$= \lim_{r \to 0} \frac{re^{i\alpha}\left(\frac{d}{dz} w_1\right) w_2 + w_1 re^{i\alpha}\left(\frac{d}{dz} w_2\right) + r^2 e^{i2\alpha}\left(\frac{d}{dz} w_1\right)\left(\frac{d}{dz} w_2\right)}{re^{i\alpha}}$$

$$= \left(\frac{d}{dz} w_1\right) w_2 + w_1 \left(\frac{d}{dz} w_2\right) + \lim_{r \to 0}\left\{ re^{i\alpha}\left(\frac{d}{dz} w_1\right)\left(\frac{d}{dz} w_2\right) \right\}.$$

Since the third term vanishes, we have

$$\frac{d}{dz}(w_1 w_2) = \left(\frac{d}{dz} w_1\right) w_2 + w_1 \left(\frac{d}{dz} w_2\right). \tag{13.14}$$

Thus, if the functions $\frac{d}{dz} w_1$ and $\frac{d}{dz} w_2$ are not functions of α, the function $\frac{d}{dz}(w_1 w_2)$ is not either, which proves the theorem. ∎

We see that relation (13.14) plays a fundamental role in analysis, which justifies referring to it by a name:

Definition 53 *Let D be an operator whose domain is any algebra $\mathcal{A}$ and whose range is any linear space $\mathcal{B}$,*

$$D : \mathcal{A} \to \mathcal{B}.$$

If the relation

$$D\left(PQ\right) = \left(DP\right)Q + P\left(DQ\right) \tag{13.15}$$

*holds for all elements $P, Q \in \mathcal{A}$, we refer to it as the **Leibniz identity**.*[3]

As an application of Definition 51, let us test for analyticity the class of functions $w\left(z\right) = z^s$, where $s \in \mathbb{R}$:

$$\frac{d}{dz}\left(z^s\right) = \lim_{r \to 0}\frac{\left(z + re^{i\alpha}\right)^s - z^s}{re^{i\alpha}}$$

$$= \lim_{r \to 0}\frac{z^s + \sum_{k=1}^{\infty}\binom{s}{k}r^k e^{ik\alpha}z^{s-k} - z^s}{re^{i\alpha}}$$

$$= sz^{s-1}.$$

Since the phase factors $e^{ik\alpha}$ and $e^{i\alpha}$ cancel out in the limit $r \to 0$, the functions $\mathcal{D}\left(z^s\right)$ are analytic for all $s \in \mathbb{R}$ — though it is singular at $z = 0$ for all $s < 1$ other than zero.

The simplest example of a non-analytic function is the complex conjugate, $w\left(z\right) = z^*$:

$$\frac{d}{dz}z^* = \lim_{r \to 0}\frac{\left(z + re^{i\alpha}\right)^* - z^*}{re^{i\alpha}} = \lim_{r \to 0}\frac{re^{-i\alpha}}{re^{i\alpha}} \equiv e^{-i2\alpha}. \tag{13.16}$$

It follows that a function of z is not analytic if it involves the operator of complex conjugation.

To translate the Definition 51 of analyticity into a formal test of analyticity, we compute the derivative of a function $w\left(z\right)$ in a given direction α. Writing

$$dx + idy = \left(\cos\alpha + i\sin\alpha\right)dt = e^{i\alpha}dt$$

---

[3] The expression $(fg)^{(n)} = \sum\binom{n}{k}f^{(k)}g^{(n-k)}$ for the n-th derivative of a product of functions is known as the *Leibniz formula*. The *Leibniz identity* is its special case of $n = 1$.

for an auxiliary parameter t, we have

$$\frac{d}{dz}w \equiv \frac{du + idv}{dx + idy} = e^{-i\alpha}\left\{\cos\alpha\frac{\partial u + i\partial v}{\partial x} + \sin\alpha\frac{\partial u + i\partial v}{\partial y}\right\}.$$

Expressing $\cos\alpha$ and $\sin\alpha$ in terms of exponential functions and rearranging the terms, one obtains

$$\frac{d}{dz}w = \frac{1}{2}\left(\frac{\partial u}{\partial x} + i\frac{\partial v}{\partial x} - i\frac{\partial u}{\partial y} + \frac{\partial v}{\partial y}\right) + \frac{e^{-i2\alpha}}{2}\left(\frac{\partial u}{\partial x} + i\frac{\partial v}{\partial x} + i\frac{\partial u}{\partial y} - \frac{\partial v}{\partial y}\right).$$

The definition of analyticity (13.10) requires the vanishing of the second expression in parentheses. This yields two real equations,

$$\frac{\partial u}{\partial x} - \frac{\partial v}{\partial y} = \frac{\partial u}{\partial y} + \frac{\partial v}{\partial x} = 0, \tag{13.17}$$

which are known as the **Cauchy-Riemann** equations. If they are satisfied, the derivation operator simplifies to:

$$\frac{d}{dz} = \frac{1}{2}\left(\frac{\partial}{\partial x} - i\frac{\partial}{\partial y}\right). \tag{13.18}$$

Hence, as expected:

$$\frac{d}{dz}z = \frac{1}{2}\left(\frac{\partial}{\partial x} - i\frac{\partial}{\partial y}\right)(x + iy) = 1, \tag{13.19}$$

$$\frac{d}{dz^*}z = \frac{1}{2}\left(\frac{\partial}{\partial x} + i\frac{\partial}{\partial y}\right)(x + iy) = 0. \tag{13.20}$$

We shall now reformulate these results in a more abstract form.

A conceptual approach to analyticity

In the transition from real to complex numbers, the geometric interpretation of the derivative is not preserved, but the defining formula (13.1) remains valid in the form (13.8). In the transition from complex numbers to quantions, the defining formula (13.8) is not preserved either. Thus, if a quantionic derivative exists, it is not a generalization of the formula (13.1). We are thus to seek other characterizations that are mutually equivalent in the complex domain, but not necessarily equivalent in the quantionic domain. The following observations suggest a characterization we shall call "conceptual analyticity".

First observation: The definition (13.8) and the condition (13.10) are postulates, while relations (13.19) and (13.18) are theorems. Yet, these theorems are conceptually simpler than the postulates. This suggests that they should be selected as the postulates.

Second observation: The relation $\frac{d}{dz}z = 1$ tells us that z is the independent variable, or, in the language of algebra, the generator of the algebra of analytic functions.

Third observation: The relation $\frac{d}{dz}z^* = 0$ (equivalent to $\frac{d}{dz^*}z = 0$) means that the complex conjugate of the independent variable is a constant with respect to the derivation operator $\frac{d}{dz}$.

By postulating relations (13.19) and (13.20), the Cauchy-Riemann equations (13.17) and the expression (13.18) for the derivation operator follow as theorems. To prove it, we note that the most general homogeneous first order differential operator in $\mathbb{C}$ is a linear combination of two partial derivatives,

$$\frac{d}{dz} = \left(r\frac{\partial}{\partial x} + s\frac{\partial}{\partial y} \right),$$

where r and s are unknown complex coefficients. The conditions (13.19) and (13.20) imply, respectively:

$$\frac{d}{dz}z = \left(r\frac{\partial}{\partial x} + s\frac{\partial}{\partial y} \right)(x + iy) = r + is = 1,$$

$$\frac{d}{dz}z^* = \left(r\frac{\partial}{\partial x} + s\frac{\partial}{\partial y} \right)(x - iy) = r - is = 0.$$

The solutions

$$r = \frac{1}{2} \quad \text{and} \quad s = -ir$$

yield the expression (13.18) for $\frac{d}{dz}$.

Theorem 54 *The conceptual form of the Cauchy-Riemann equations is*

$$\frac{d}{dz^*}w \equiv \frac{d}{dz}w^* = 0. \tag{13.21}$$

Proof. Substitution of the expressions (13.18) for $\frac{d}{dz}$ and of (13.9) for w yields

$$
\begin{aligned}
2\frac{d}{dz^*}w &= \left(\frac{\partial}{\partial x} + i\frac{\partial}{\partial y}\right)(u + iv) \\
&= \left(\frac{\partial u}{\partial x} - \frac{\partial v}{\partial y}\right) + i\left(\frac{\partial v}{\partial x} + \frac{\partial u}{\partial y}\right) = 0,
\end{aligned}
$$

which is equivalent to (13.17). ∎

Two essential properties of analytic functions are given by the following theorems:

Theorem 55 (Stability) *The derivative of an analytic function is an analytic function at all values of z at which $\frac{d}{dz}$ and $\frac{d}{dz^*}$ commute.*

Proof. By definition, the function $w(z)$ is analytic if $\frac{d}{dz^*}w = 0$. This implies $\frac{d}{dz}\frac{d}{dz^*}w = 0$. If $\frac{d}{dz}$ and $\frac{d}{dz^*}$ commute at the point z, we may write $\frac{d}{dz^*}\frac{d}{dz}w \equiv \frac{d}{dz^*}\left(\frac{d}{dz}w\right) = 0$, which, by the definition of analyticity, implies that the function $\frac{d}{dz}w$ is analytic. ∎

Theorem 56 (Harmonicity) *Analytic functions w are harmonic functions, that is,*

$$\Delta w \equiv 0,$$

at all values of z at which $\frac{d}{dz}$ and $\frac{d}{dz^*}$ commute.

Proof. At all points z where

$$\frac{\partial}{\partial x}\frac{\partial}{\partial y}w(z) = \frac{\partial}{\partial y}\frac{\partial}{\partial x}w(z),$$

we have

$$
\begin{aligned}
\frac{d}{dz}\frac{d}{dz^*} &= \frac{d}{dz^*}\frac{d}{dz} = \frac{1}{2}\left(\frac{\partial}{\partial x} - i\frac{\partial}{\partial y}\right)\frac{1}{2}\left(\frac{\partial}{\partial x} + i\frac{\partial}{\partial y}\right) \\
&= \frac{1}{4}\left(\frac{\partial^2}{\partial x^2} + \frac{\partial^2}{\partial y^2}\right) = \frac{1}{4}\Delta,
\end{aligned}
$$

which proves the assertion. ∎

This completes the set of basic properties of analytic functions. They are listed for reference in the following table.

Linearity	
Stability	Theorem 55
Leibniz identity	Definition 53
Existence of a generator	Relation (13.19)
Analyticity	Equations (13.20)
Harmonicity	Theorem 56

13.2   Analyticity in the matrix formalism

While the defining formula (13.7) is not applicable to quantions, the conceptual approach to complex analysis is applicable in principle. Practically, it consists in formulating the six tabulated properties in the matrix formalism based on relation (2.38)

The transition from the standard expression $z = x + iy$ to its matrix form (see page 74) is effected by substituting the identity matrix I and the matrix J (which represents a rotation by $\pi/2$ in the Gaussian plane) for the real and imaginary units 1 and i :

$$Z = xI + yJ = \begin{pmatrix} x & -y \\ y & x \end{pmatrix}. \tag{13.22}$$

Hence, the matrix form of a function $w(z) = u(x, y) + iv(x, y)$ is

$$W(Z) = u(x, y) I + v(x, y) J = \begin{pmatrix} u & -v \\ v & u \end{pmatrix}. \tag{13.23}$$

We shall now construct a matrix $\mathcal{D}_c$ which has all the listed properties of derivation operators.

Linearity is guaranteed if $\mathcal{D}_c$ is represented by a matrix

$$\mathcal{D}_c \overset{def}{=} \begin{pmatrix} \alpha & \gamma \\ \beta & \delta \end{pmatrix} \tag{13.24}$$

whose elements are to be determined from the above conditions.

Stability requires that the matrix

$$\mathcal{D}_c W = \begin{pmatrix} \alpha & \gamma \\ \beta & \delta \end{pmatrix} \begin{pmatrix} u & -v \\ v & u \end{pmatrix} = \begin{pmatrix} \alpha u + \gamma v & \gamma u - \alpha v \\ \beta u + \delta v & \delta u - \beta v \end{pmatrix}$$

be of the form (13.23), which implies the following identities:

$$\alpha u + \gamma v \; \equiv \; \delta u - \beta v,$$
$$\beta u + \delta v \; \equiv \; -(\gamma u - \alpha v) = \alpha v - \gamma u.$$

Hence, $\delta = \alpha$ and $\beta = -\gamma$, which yields

$$\mathcal{D} = \begin{pmatrix} \alpha & \gamma \\ -\gamma & \alpha \end{pmatrix}. \tag{13.25}$$

To impose the *Leibniz identity* (13.15) on two functions

$$P(x,y) = \begin{pmatrix} p & -q \\ q & p \end{pmatrix}, \quad Q(x,y) = \begin{pmatrix} u & -v \\ v & u \end{pmatrix},$$

we compute the two sides of (13.15) separately. Ignoring stability, we shall work with the operator (13.24).

The left-hand side is:

$$L.H.S.$$
$$= \begin{pmatrix} \alpha & \gamma \\ \beta & \delta \end{pmatrix} \begin{pmatrix} p & -q \\ q & p \end{pmatrix} \begin{pmatrix} u & -v \\ v & u \end{pmatrix}$$
$$= \begin{pmatrix} \alpha & \gamma \\ \beta & \delta \end{pmatrix} \begin{pmatrix} pu - qv & -pv - qu \\ pv + qu & pu - qv \end{pmatrix}$$
$$= \begin{pmatrix} \alpha(pu - qv) + \gamma(pv + qu) & -\alpha(pv + qu) + \gamma(pu - qv) \\ \beta(pu - qv) + \delta(pv + qu) & -\beta(pv + qu) + \delta(pu - qv) \end{pmatrix}$$

The right-hand side is:

$$R.H.S.$$
$$= \left\{ \begin{pmatrix} \alpha & \gamma \\ \beta & \delta \end{pmatrix} \begin{pmatrix} p & -q \\ q & p \end{pmatrix} \right\} \begin{pmatrix} u & -v \\ v & u \end{pmatrix} + \begin{pmatrix} p & -q \\ q & p \end{pmatrix} \begin{pmatrix} \alpha & \gamma \\ \beta & \delta \end{pmatrix} \begin{pmatrix} u & -v \\ v & u \end{pmatrix}$$
$$= \begin{pmatrix} (\alpha p + \gamma q)u + (\gamma p - \alpha q)v & -(\alpha p + \gamma q)v + (\gamma p - \alpha q)u \\ (\beta p + \delta q)u + (\delta p - \beta q)v & -(\beta p + \delta q)v + (\delta p - \beta q)u \end{pmatrix}$$
$$+ \begin{pmatrix} p(\alpha u + \gamma v) - q(\beta u + \delta v) & p(\gamma u - \alpha v) - q(\delta u - \beta v) \\ q(\alpha u + \gamma v) + p(\beta u + \delta v) & q(\gamma u - \alpha v) + p(\delta u - \beta v) \end{pmatrix}$$

Hence, the Leibniz identity for the matrix element [11] reads:

$$\alpha(pu - qv) + \gamma(pv + qu)$$
$$\equiv (\alpha p + \gamma q)u + (\gamma p - \alpha q)v + p(\alpha u + \gamma v) - q(\beta u + \delta v).$$

Taking in turn $q = v = 0$, $q = u = 0$, $p = v = 0$, and $p = u = 0$, one obtains the four identities

$$\alpha (pu) \equiv (\alpha p) u + p (\alpha u), \tag{13.26}$$
$$\gamma (pv) \equiv (\gamma p) v + p (\gamma v), \tag{13.27}$$
$$\gamma (qu) \equiv (\gamma q) u - q (\beta u), \tag{13.28}$$
$$-\alpha (qv) \equiv - (\alpha q) v - q (\delta v). \tag{13.29}$$

The matrix elements [12], [21], and [22] yield the same results.
   The two identities (13.28) and (13.29) yield

$$\delta = \alpha,$$
$$\beta = -\gamma,$$

which means that the Leibniz identity implies the stability conditions.
   The two identities (13.26) and (13.27) are Leibniz identities for α and γ. Hence, the most general solutions are homogeneous first order linear differential operators,

$$\alpha = a\frac{\partial}{\partial x} + b\frac{\partial}{\partial y},$$
$$\gamma = c\frac{\partial}{\partial x} + d\frac{\partial}{\partial y},$$

where the real coefficients a, b, c, d remain to be determined.
   These results restrict the operator $\mathcal{D}$ to the form

$$\mathcal{D}_c = \begin{pmatrix} a\partial_x + b\partial_y & c\partial_x + d\partial_y \\ -c\partial_x - d\partial_y & a\partial_x + b\partial_y \end{pmatrix}. \tag{13.30}$$

We can also obtain this solution by a more conceptual procedure which is directly applicable to quantions.
   To this end, consider the special case of a constant matrix P, so that $(\mathcal{D}_c P) = 0$. The Leibniz identity (13.15) reduces to

$$\mathcal{D}_c (PQ) \equiv P (\mathcal{D}_c Q),$$

which implies that the matrices $\mathcal{D}_c$ and P must commute:

$$\begin{pmatrix} \alpha & \gamma \\ \beta & \delta \end{pmatrix} \begin{pmatrix} p & -q \\ q & p \end{pmatrix} - \begin{pmatrix} p & -q \\ q & p \end{pmatrix} \begin{pmatrix} \alpha & \gamma \\ \beta & \delta \end{pmatrix} \equiv 0.$$

After simplification, this yields

$$\begin{pmatrix} \beta + \gamma & \delta - \alpha \\ \delta - \alpha & -\beta - \gamma \end{pmatrix} q \equiv 0,$$

where the identity means 'for every q'. Hence, $\delta = \alpha$ and $\beta = -\gamma$, which confirms the previous result. The Leibniz identity now reads

$$\begin{pmatrix} \alpha & \gamma \\ -\gamma & \alpha \end{pmatrix} \left\{ \begin{pmatrix} p & -q \\ q & p \end{pmatrix} \begin{pmatrix} u & -v \\ v & u \end{pmatrix} \right\}$$

$$\equiv \left\{ \begin{pmatrix} \alpha & \gamma \\ -\gamma & \alpha \end{pmatrix} \begin{pmatrix} p & -q \\ q & p \end{pmatrix} \right\} \begin{pmatrix} u & -v \\ v & u \end{pmatrix}$$

$$+ \begin{pmatrix} p & -q \\ q & p \end{pmatrix} \left\{ \begin{pmatrix} \alpha & \gamma \\ -\gamma & \alpha \end{pmatrix} \begin{pmatrix} u & -v \\ v & u \end{pmatrix} \right\}.$$

To verify it, it suffices to compute a single term of the expansion, let's say [11]:

$$\alpha \, (pu) - \alpha \, (qv) + \gamma \, (pv) + \gamma \, (qu)$$

$$\equiv \ (\alpha p) \, u + (\gamma q) \, u + (\gamma p) \, v - (\alpha q) \, v$$

$$+ p \, (\alpha u) + p \, (\gamma v) - q \, (\alpha v) + q \, (\gamma u) .$$

This implies that the matrix elements α and γ must also satisfy the Leibniz identity

$$\alpha \, (up) \ \equiv \ (\alpha u) \, p + u \, (\alpha p) ,$$

$$\gamma \, (vp) \ \equiv \ (\gamma v) \, p + v \, (\gamma p) ,$$

which implies the solution (13.30).

The *generator* (13.22) of the matrix algebra (the independent variable) must satisfy the condition $DZ = I$. Explicitly:

$$\begin{pmatrix} a\partial_x + b\partial_y & c\partial_x + d\partial_y \\ -c\partial_x - d\partial_y & a\partial_x + b\partial_y \end{pmatrix} \begin{pmatrix} x & -y \\ y & x \end{pmatrix} = \begin{pmatrix} a+d & -b+c \\ -c+b & d+a \end{pmatrix} = \begin{pmatrix} 1 & 0 \\ 0 & 1 \end{pmatrix}$$

Hence, for an arbitrary real parameter r :

$$a \ = \ \frac{1}{2} + r,$$

$$d \ = \ \frac{1}{2} - r,$$

$$c \ = \ b.$$

The derivation operator $\mathcal{D}_c$ now contains only two unknown parameters, r and b :

$$\mathcal{D}_c = \begin{pmatrix} \left(\frac{1}{2}+r\right)\partial_x + b\partial_y & b\partial_x + \left(\frac{1}{2}-r\right)\partial_y \\ -b\partial_x - \left(\frac{1}{2}-r\right)\partial_y & \left(\frac{1}{2}+r\right)\partial_x + b\partial_y \end{pmatrix}. \tag{13.31}$$

The *Cauchy-Riemann* equation (13.20), applied to $W = Z$, reads

$$
\begin{aligned}
\mathcal{D}_c^\dagger Z &= \begin{pmatrix} \left(\frac{1}{2}+r\right)\partial_x + b\partial_y & -b\partial_x - \left(\frac{1}{2}-r\right)\partial_y \\ b\partial_x + \left(\frac{1}{2}-r\right)\partial_y & \left(\frac{1}{2}+r\right)\partial_x + b\partial_y \end{pmatrix} \begin{pmatrix} x & -y \\ y & x \end{pmatrix} \\
&= \begin{pmatrix} \left(\frac{1}{2}+r\right) - \left(\frac{1}{2}-r\right) & -b-b \\ b+b & -\left(\frac{1}{2}-r\right) + \left(\frac{1}{2}+r\right) \end{pmatrix} = 0,
\end{aligned}
$$

which implies $r = b = 0$. Thus, the operator

$$\mathcal{D}_c = \frac{1}{2}\begin{pmatrix} \partial_x & \partial_y \\ -\partial_y & \partial_x \end{pmatrix} \tag{13.32}$$

agrees with the standard form (13.18).

The conceptual *Cauchy-Riemann* equation (13.20), applied to an arbitrary function W, reads

$$\mathcal{D}^\dagger W = 0, \tag{13.33}$$

or, after expansion,

$$\begin{pmatrix} \partial_x & -\partial_y \\ \partial_y & \partial_x \end{pmatrix}\begin{pmatrix} u & -v \\ v & u \end{pmatrix} = \begin{pmatrix} \partial_x u - \partial_y v & -\partial_x v - \partial_y u \\ \partial_y u + \partial_x v & -\partial_y v + \partial_x u \end{pmatrix} = 0.$$

This yields the standard Cauchy-Riemann equations (13.17).

Harmonicity, Theorem 56, is readily verified:

$$
\begin{aligned}
\mathcal{D}^\dagger \mathcal{D} &= \frac{1}{2}\begin{pmatrix} \partial_x & -\partial_y \\ \partial_y & \partial_x \end{pmatrix}\frac{1}{2}\begin{pmatrix} \partial_x & \partial_y \\ -\partial_y & \partial_x \end{pmatrix} \\
&= \frac{1}{4}\begin{pmatrix} \frac{\partial^2}{\partial x^2} + \frac{\partial^2}{\partial y^2} & 0 \\ 0 & \frac{\partial^2}{\partial x^2} + \frac{\partial^2}{\partial y^2} \end{pmatrix} = \frac{1}{4}I\Delta. \tag{13.34}
\end{aligned}
$$

We see that the matrix formalism encapsulates the essential aspects of analytic functions without carrying any additional baggage from the differential calculus (like taking limits of the type $h \to 0$.)

The case of quaternions

For comparative purposes, let us consider the case of quaternions. A quaternionic function in matrix form is

$$Q = \begin{pmatrix} W & -X & -Y & -Z \\ X & W & Z & -Y \\ Y & -Z & W & X \\ Z & Y & -X & W \end{pmatrix},$$

where W, X, Y, Z are real functions of four real variables w, x, y, z. Equating in turn pairs of off-diagonal elements to zero yields reducible 4×4 matrix representations of complex numbers. For example, taking $X = Z = 0$ simplifies Q to

$$Q = W \begin{pmatrix} 1 & 0 & 0 & 0 \\ 0 & 1 & 0 & 0 \\ 0 & 0 & 1 & 0 \\ 0 & 0 & 0 & 1 \end{pmatrix} + Y \begin{pmatrix} 0 & 0 & -1 & 0 \\ 0 & 0 & 0 & -1 \\ 1 & 0 & 0 & 0 \\ 0 & 1 & 0 & 0 \end{pmatrix}.$$

This brings us back into the complex domain in reducible form

$$\mathcal{D}_c = \begin{pmatrix} \partial_w & * & \partial_y & * \\ * & \partial_w & * & \partial_y \\ -\partial_y & * & \partial_w & * \\ * & -\partial_y & * & \partial_w \end{pmatrix}.$$

Two additional steps yield the complete solution:

$$\mathcal{D}_q = \begin{pmatrix} \partial_w & \partial_x & \partial_y & \partial_z \\ -\partial_x & \partial_w & -\partial_z & \partial_y \\ -\partial_y & \partial_z & \partial_w & -\partial_x \\ -\partial_z & -\partial_y & \partial_x & \partial_w \end{pmatrix}. \qquad (13.35)$$

The matrix $\mathcal{D}_q$ is thus structurally a quaternion. Since quaternions do not commute, the matrix $\mathcal{D}_q$ does not satisfy the Leibniz identity.[4]

---

[4] There are also other approaches to quaternionic analysis, but they have not been found applicable to quantions (at least, not by the present author). Example are: A. Sudbery: *Quaternionic Analysis, Math. Proc. Camb. Phil. Soc.* **85**, 199-225 (1979), and Vladislav V. Kravchenko: *Applied Quaternionic Analysis,* Heldermann Verlag, Lemgo, Germany (2003).

13.3 Derivation in the algebra of quantions

To the matrix expression (13.22) for a complex independent variable corresponds the expression (2.18) for a hypothetical quantionic independent variable. Similarly, to the expression (13.23) for a complex function correspond the expression

$$Q = w^0 \Lambda_0 + w^1 \Lambda_1 + w^2 \Lambda_2 + w^3 \Lambda_3 \qquad (13.36)$$

for quantionic functions, where the coefficients w^μ are functions of the four complex variables, that is $w^\mu = w^\mu \left(z^0, z^1, z^2, z^3 \right)$. In the linear subspace $\mathcal{L}^h$ of Hermitian quantions, the four independent variables x^μ and the four functions $u^\mu = u^\mu \left(x^0, x^1, x^2, x^3 \right)$ are real.

These objects are all collected in the following diagram, in which the horizontal arrows represent restrictions from the algebra of quantions to the real and complex fields, while the vertical arrows represent restrictions from complex to real structures:

$\mathbb{C}$
$Z = xI + yJ$
$W = uI + vJ$

$\longleftarrow$

$\mathcal{L}$
$Z = z^0 I + z^1 \Lambda_1 + z^2 \Lambda_2 + z^3 \Lambda_3$
$Q = w^0 I + w^1 \Lambda_1 + w^2 \Lambda_2 + w^3 \Lambda_3$

$\downarrow$ $\downarrow$

$\mathbb{R}$
$X = xI$
$U = u(x)I$

$\longleftarrow$

$\mathcal{L}^h$
$X = x^0 I + x^1 \Lambda_1 + x^2 \Lambda_2 + x^3 \Lambda_3$
$H = u^0 I + u^1 \Lambda_1 + u^2 \Lambda_2 + x^3 \Lambda_3$

Given the extension procedure from (13.22) to

$$Z = \begin{pmatrix} z^0 + z^3 & z^1 - iz^2 & 0 & 0 \\ z^1 + iz^2 & z^0 - z^3 & 0 & 0 \\ 0 & 0 & z^0 + z^3 & z^1 - iz^2 \\ 0 & 0 & z^1 + iz^2 & z^0 - z^3 \end{pmatrix}, \qquad (13.37)$$

the next task is to extend the complex derivation operator (13.32) to quantions. To this end, we shall need the partial derivation operators with respect to the complex variables z^0 to z^3 :

$$\nabla_\mu \overset{def}{=} \frac{\partial}{\partial z^\mu}.$$

Clearly, the quantionic derivation operator $\mathcal{D}$ must be a 4×4 matrix whose elements are linear combinations of the operators ∇_μ. The only concrete guidelines for constructing it are the five properties of complex analysis listed on page 13.1.

Stability: Corresponding to the proof of (13.25), stability requires that the derivation operator be formally an L-type quantion,

$$\mathcal{D} = \begin{pmatrix} \alpha & \gamma & 0 & 0 \\ \beta & \delta & 0 & 0 \\ 0 & 0 & \alpha & \gamma \\ 0 & 0 & \beta & \delta \end{pmatrix} \in \mathcal{L}, \qquad (13.38)$$

but the non-commutativity of quantions implies that this operator does not satisfy the Leibniz condition.

The *Leibniz identity:* We can satisfy the Leibniz identity (13.15) by taking the matrix $\mathcal{D}$ in the commutant of $\mathcal{L}$:

$$\mathcal{D} = \begin{pmatrix} \alpha & 0 & \gamma & 0 \\ 0 & \alpha & 0 & \gamma \\ \beta & 0 & \delta & 0 \\ 0 & \beta & 0 & \delta \end{pmatrix} \in \mathcal{R}. \qquad (13.39)$$

This operator satisfies the identity (13.15) for quantions if the elements α, β, γ, and δ are partial differential operators, but since

$$\mathcal{D}Q = \begin{pmatrix} \alpha u & \alpha w & \gamma u & \gamma w \\ \alpha v & \alpha z & \gamma v & \gamma z \\ \beta u & \beta w & \omega u & \omega w \\ \beta v & \beta z & \omega v & \omega z \end{pmatrix} \notin \mathcal{L}, \qquad (13.40)$$

it does not satisfy the stability property.

We are now to chose one of the following options:

(1) Impose stability and drop the Leibniz identity — in which case $\mathcal{D}$ would be given by (13.38).

(2) Impose the Leibniz identity and drop stability — in which case $\mathcal{D}$ would be given by (13.39).

We shall 'borrow' the answer from Part II, where Zovko's interpretation and structural quantization yield a system of four complex field

equations as a theorem. These equations can be compactly formulated as a single quantionic differential equation, in which the partial
derivation operators are encapsulated in a quantionic derivation operator. The encapsulation is unique and of the type (13.39). We thus
conclude that *the Leibniz identity takes precedence over stability.*

This brings up the question of what could be the use of a deriva-
tion operator $\mathcal{D}$ if, for every quantionic field Q, the matrix $Q + \varepsilon \mathcal{D}Q$
in its infinitesimal neighborhood is not a quantionic field? The an-
swer to this question — also suggested by the physical quantionic field
equations — is given on page 170.

The *generator:* By declaring Z in (13.37) as the independent quantionic variable, the condition $\mathcal{D}Z = I$ imposes strong defining conditions on the differential operators α to δ in the matrix (13.39):

$$\mathcal{D}Z = \begin{pmatrix} \alpha a & \alpha c & \gamma a & \gamma c \\ \alpha b & \alpha d & \gamma b & \gamma d \\ \beta a & \beta c & \delta a & \delta c \\ \beta b & \beta d & \delta b & \delta d \end{pmatrix} = \begin{pmatrix} 1 & 0 & 0 & 0 \\ 0 & 1 & 0 & 0 \\ 0 & 0 & 1 & 0 \\ 0 & 0 & 0 & 1 \end{pmatrix},$$

where a, b, c, d represent the variables in (13.37). These conditions are

$$\frac{\partial a}{\partial a} = \frac{\partial b}{\partial b} = \frac{\partial c}{\partial c} = \frac{\partial d}{\partial d} = 1,$$
$$\frac{\partial a}{\partial b} = \cdots = \frac{\partial d}{\partial c} = 0.$$

Hence, the solutions for the operators α to δ are

$$\alpha = \delta = \partial_a + \partial_d,$$
$$\beta = \gamma = 0.$$

They trivialize the matrix $\mathcal{D}$ to the scalar operator

$$\mathcal{D} = (\partial_a + \partial_d),$$

which ignores the variables b, c, and $a - d$. Hence, a nontrivial derivation operator $\mathcal{D}$ and an independent variable Z cannot exist simultaneously. It follows that if we keep $\mathcal{D}$, the algebra of matrices (13.36)
is not a generated algebra. This means that the quantionic functions

$Q\left(z^0, z^1, z^2, z^3\right)$ over which the Leibniz identity is satisfied cannot be written in the form of polynomials or power series in$\in Z$.

If we restrict the quantions to the complex numbers by taking $z^1 = z^2 = z^3 = 0$, the derivation operator $\mathcal{D}$ must yield the operator $\mathcal{D}_c$ given by (13.32). This is the case if and only if $\mathcal{D}$ is of the form

$$\mathcal{D} = I\frac{\partial}{\partial z^0} + \mathcal{D}_0,$$

where $\mathcal{D}_0$ is traceless and does not contain $\frac{\partial}{\partial z^0}$, $\frac{\partial}{\partial z^2}$ and $\frac{\partial}{\partial z^3}$. Hence:

$$\mathcal{D} = \begin{pmatrix} (\nabla_0 + \nabla_3)\,I & \gamma \\ \beta & (\nabla_0 - \nabla_3)\,I \end{pmatrix}.$$

There are two nontrivial solutions:

$$\mathcal{D} = \begin{pmatrix} (\nabla_0 + \nabla_3)\,I & (\nabla_1 \pm i\nabla_2) \\ (\nabla_1 \mp i\nabla_2)\,I & (\nabla_0 - \nabla_3)\,I \end{pmatrix}.$$

We expect $\mathcal{D}^h$ (the Hermitian part of $\mathcal{D}$) to be the physically relevant operator. One obtains it by substituting ∂_μ for ∇_μ. In terms of the Newman-Penrose Symbols,[5]

$$\left.\begin{aligned} D &\overset{def}{=} \partial_0 + \partial_3, \\ \delta^* &\overset{def}{=} \partial_1 - i\partial_2, \\ \delta &\overset{def}{=} \partial_1 + i\partial_2, \\ \Delta &\overset{def}{=} \partial_0 - \partial_3, \end{aligned}\right\} \tag{13.41}$$

the two solutions are

$$\left.\begin{aligned} \mathcal{D}_1 &= \begin{pmatrix} DI & \delta I \\ \delta^* I & \Delta I \end{pmatrix}, \\ \mathcal{D}_2 &= \begin{pmatrix} DI & \delta^* I \\ \delta I & \Delta I \end{pmatrix} \equiv \mathcal{D}_1^*. \end{aligned}\right\} \tag{13.42}$$

Analyticity: If the quantionic functions are defined only over the real Minkowski space, the concept of analyticity is not applicable. It

---

[5] Newman, E. T. and Penrose, R. "An Approach to Gravitational Radiation by a Method of Spin Coefficients". J. Math. Phys. **3**, 566 (1962).

is conceivable that future insights will modify this conclusion, but this would require some new ideas that cannot be guessed at this point.

Harmonicity: In relation (13.34), the operator $A\left(\mathcal{D}_c\right) = \mathcal{D}_c^\dagger\mathcal{D}_c$ is the Laplacian. Since the complex numbers are degenerate, we have $\mathcal{D}_c^\dagger = \mathcal{D}_c^\#$, so that $M\left(\mathcal{D}_c\right) = \mathcal{D}_c^\#\mathcal{D}_c$ is the Laplacian as well. It is only the second version that generalizes to quantions:

$$
\begin{aligned}
\mathcal{D}_1^\#\mathcal{D}_1 &= \begin{pmatrix} \Delta I & -\delta I \\ -\delta^* I & DI \end{pmatrix} \begin{pmatrix} DI & \delta I \\ \delta^* I & \Delta I \end{pmatrix} \\
&= \left(\Delta D - \delta\delta^*\right) I \\
&= \left[\left(\partial_0 - \partial_3\right)\left(\partial_0 + \partial_3\right) - \left(\partial_1 - i\partial_2\right)\left(\partial_1 + i\partial_2\right)\right] I \\
&= \left(\partial_0^2 - \partial_3^2 - \partial_1^2 - \partial_2^2\right) I.
\end{aligned}
$$

The same is true for the solution $\mathcal{D}_2$. Thus: *The metric norm of the Hermitian quantionic derivation operator is the D'Alambertian:*

$$
m\left(\mathcal{D}_1\right) = \mathcal{D}_1^\#\mathcal{D}_1 = \square. \tag{13.43}
$$

Since the Laplacian Δ, defined in the Gaussian space of complex numbers, generalizes to the D'Alambertian $\square$, defined in the Gaussian space of quantions, it is natural to extend the terminology "harmonic function" from ϕ satisfying the equation $\Delta\phi = 0$ to ϕ satisfying the equation $\square\phi = 0$. This is what is meant by the statement that the 'harmonicity condition' is satisfied in the quantionic domain.

This completes the construction of the quantionic derivation operator as far as it can be carried out at this point.

Let us tabulate the properties of the operator $\mathcal{D}$:

Linearity	YES
Stability	NO
Leibniz identity	YES
Existence of a generator	NO
Analyticity	NO
Harmonicity (for $\mathcal{D}_1$ and $\mathcal{D}_2$)	YES

We conclude with brief discussions of some properties of $\mathcal{D}_1$.

Why are there two solutions? Reminder: The algebraic norm $A(Q)$ of a quantionic field $Q(x)$ is a field of future-oriented time-like Minkowski vectors. Write it as $j^\mu(x)$.

From Part II: Interpreting j^μ as a current leads us to consider the non-linear equation of continuity $\partial_\mu j^\mu = 0$. Reducing it to a linear equation by structural quantization gives rise to the operator $\mathcal{D}_1$.

Answer: $A(Q) = Q^\dagger Q$ was conventionally taken to be the norm, but $B(Q) = QQ^\dagger$ could have been selected as well. If $B(Q)$ is interpreted as the current, the equation of continuity gives rise to $\mathcal{D}_2$.

Conclusion: Only $\mathcal{D}_1$ is relevant.

Lack of stability By "lack of stability" we mean that the derivative $\mathcal{D}Q$ of a quantionic field $Q(x) \in \mathcal{L}$ is not in $\mathcal{L}$. This implies that quantionic differential equations of motion do not exist for quantionic fields. This would be a fatal flaw if it were not for the concept of "linking", which associates a unique vector $|q\rangle \in \mathcal{H}_q$ to every matrix $Q \in \mathcal{M}$. Thus, even though the matrix $\mathcal{D}Q$ cannot appear meaningfully by itself in a field equation, it is meaningful as an operator acting in $\mathcal{H}_q$. This is due to the associativity of the algebra of matrices:

$$(\mathcal{D}Q)|\omega\rangle \equiv \mathcal{D}[Q|\omega\rangle] = \mathcal{D}|q\rangle.$$

To see that the Leibniz identity is consistent with the transfer of $\mathcal{D}$ from $\mathcal{L}$ to $\mathcal{H}_q$, we apply $\mathcal{D}$ to PQ :

$$[\mathcal{D}(PQ)]|\omega\rangle$$
$$= (\mathcal{D}P)Q|\omega\rangle + P(\mathcal{D}Q)|\omega\rangle$$
$$= (\mathcal{D}P)|q\rangle + P\mathcal{D}|q\rangle \in \mathcal{H}_q.$$

It is interesting to note that the quantionic field equation derived independently in Part II without reference to the present chapter naturally appears as an equation for $|q\rangle$, not for Q.

The interpretation of the operator $\mathcal{D}_1$. The trace of $\mathcal{D}_1$ is

$$Tr(\mathcal{D}_1) = D + \Delta = 2\frac{\partial}{\partial x^0}.$$

Since the trace is structurally distinguished, so is the the derivative $\frac{\partial}{\partial x^0}$. We may thus regard the quantionic derivation operator $\mathcal{D}_1$ as the 'quantionic version' of the time derivative $\frac{d}{dt}$ of classical physics. This view is supported by the expression for the quantionic field equation derived in Part II. This equation is of the type

$$\mathcal{D}_1 \, |q\rangle = \text{Linear function of } |q\rangle \text{ and } |q^*\rangle.$$

It is thus justified to speak indifferently of the "quantionic field equation" or of the "quantionic equation of motion".

Non-existence of a generator The concept of a generator is meaningful for mappings of an algebra onto itself. This is the case of complex analytic functions. Thus, in the matrix formalism, an analytic function $w = w(z)$ is of the form

$$\begin{pmatrix} u & -v \\ v & u \end{pmatrix} = \text{Power series in } \begin{pmatrix} x & -y \\ y & x \end{pmatrix}.$$

For this reason, $Z = \begin{pmatrix} x & -y \\ y & x \end{pmatrix}$ may be called the generator of the algebra of power series.

There is no such concept in quantum field theory because complex fields are mappings of the affine Minkowski space M^4 into $\mathbb{C}$, and M^4 is not an algebra.

In the quantionic domain, functions that map $\mathcal{L}$ into itself appear in two guises:

$$\mathcal{L} \text{ viewed as an algebra} \quad \rightarrow \quad \mathcal{L}$$
$$\mathcal{L} \text{ viewed as a Gaussian space} \quad \rightarrow \quad \mathcal{L}$$

In the first interpretation, the concept of a generator exists, but a concrete generator does not exist. At this point, this interpretation seems to play no role in physics.

In the second interpretation, the concept of a generator does not exist. This is the interpretations needed in physics because the Gaussian space of the algebra $\mathcal{L}^h$ is the real linear Minkowski space M_0^4.

Part II

THE PHYSICS OF
QUANTIONS

Chapter 14

Zovko's Interpretation and Structural Quantization

The role of interpretations in physics is to establish one-to-one correspondences between the mathematical concepts over which theories are built and the numerical values in which experimental results are reported. Interpretations are neither axioms nor theorems. They are best regarded as independent insights.

In classical mechanics, the question of interpretation does not arise (other than philosophically) because the classical theoretical concepts are in a self-evident one-to-one correspondence with the observable properties of physical systems: States are represented by distributions over phase space, and observables by smooth functions over the same space. Moreover, the time evolution of a classical physical system is not affected by observations, so that the kinematics of such a system may be closely followed and compared at every instant of time with the integration of differential equations of motion that model the dynamics of the system.

Nonrelativistic quantum mechanics

In quantum mechanics, the interpretations are far from obvious because the smooth time evolution of a quantum mechanical system cannot be followed experimentally. Schrödinger arrived at his equation — which follows this evolution theoretically — by interpreting the spectra of self-adjoint operators that act on an auxiliary 'wave function' as the only possible outcomes of measurements. The wave function itself went for over six months in search of an interpretation, some four or five having been suggested by several physicists (including Schrödinger) before Born discovered the interpretation that has since been at the foundation of nonrelativistic quantum mechanics. Schrödinger's equation and Born's interpretation, taken together, imply the equation of continuity for a probability density viewed as a conserved freely compressible fluid.

The fundamental postulate of nonrelativistic quantum mechanics is the **nonrelativistic canonical quantization** procedure

$$\left.\begin{array}{c} \vec{r} \rightarrow \vec{r}, \\ t \rightarrow t, \\ \vec{p} \rightarrow -i\hbar\nabla, \\ E \rightarrow i\hbar\frac{\partial}{\partial t}. \end{array}\right\} \qquad (14.1)$$

The canonical quantization of the classical total energy

$$E = \frac{1}{2m}p^2 + V(\vec{r}, t)$$

yields the Schrödinger equation. By Born's interpretation, the phase factor $e^{i\chi} \in U(1)$ of the complex wave function $\psi = re^{i\chi}$ is not observable, while the algebraic norm

$$A(\psi) = r^2 = \rho \in \mathbb{R}$$

represents a probability density. The self-consistency of this idea is guaranteed by the existence of a vector $\vec{j}$, defined by the wave function ψ, which, together with the density function $\rho = \psi^*\psi$, satisfies the equation of continuity

$$\dot{\rho} + \nabla \cdot \vec{j} = 0. \qquad (14.2)$$

These ideas are illustrated in the first diagram in Figure 14.3. on page 180, where the double arrows represent logical implications.

Historically, Born's interpretation was a feat of intuition, but its subsequent justification is to be found in the agreement of its consequences with observations. The first such justification is that it leads to the classical equation of continuity for the probability density.

Relativistic quantum mechanics

As illustrated in the second diagram in Figure 14.3, Dirac's relativistic field equation has been obtained in two steps. In the first step — which is analogous to the first step in the nonrelativistic theory — the relativistic canonical quantization rules

$$x^\mu \quad \rightarrow \quad x^\mu,$$
$$p_\mu \quad \rightarrow \quad i\hbar\partial_\mu,$$

yield the Klein-Gordon equation, which is of second order in derivatives. In the second step, this second order is reduced to first order at the cost of additional numbers of freedom (four spin components instead of one wave function). It is interesting to note that interpretations play no role in the derivation of Dirac's equation.

14.1   Zovko's interpretation

The heuristics that led to Schrödinger's and Dirac's equations are not available in the quantionic approach to physics. A new source of differential equations was suggested in 2002 by Nikola Zovko in an oral communication. Referred to as Zovko's interpretation, it generalizes Born's interpretation of the norm of complex numbers to the algebraic norm of quantions. This is illustrated in Figure 14.1, which is adapted from Figure 11.1 on page 126. Unlike Born's interpretation, which came after Schrödinger's equation, Zovko's interpretation of quantionic fields comes first. Taken as a postulate, it yields Schrödinger's and Dirac's equations as theorems.

Born's and Zovko's interpretations are comparatively shown in Figure 11.1.

Born's interpretation states that the phase factor $e^{i\phi}$ in the factorization $\psi = e^{i\phi}r$ is not observable and that the square of the modulus represents a probability density, $\rho = r^2 \equiv \psi^*\psi$. This is possible because r^2 is a non-negative real number.

Zovko's interpretation states that the phase factor E in the factorization $Q = ER$ is not observable and that the square of the modulus represents a current of probability density, $j = R^2 \equiv Q^\dagger Q$. This is possible because R^2 is a future-oriented time-like Minkowski vector.

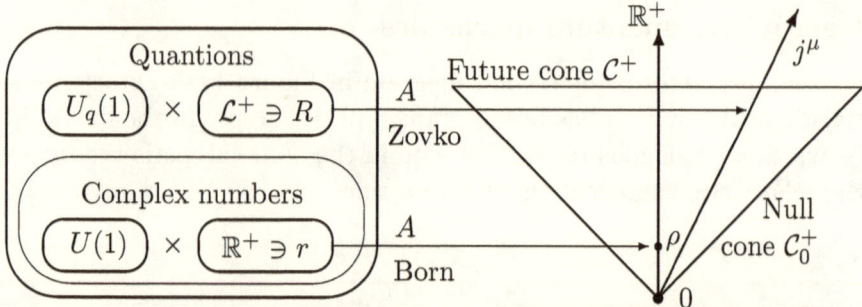

Figure 14.1: Comparison of Born's and Zovko's interpretations.

Since Zovko's interpretation plays a crucial role in what follows, it is also illustrated in the following self-explanatory diagram.

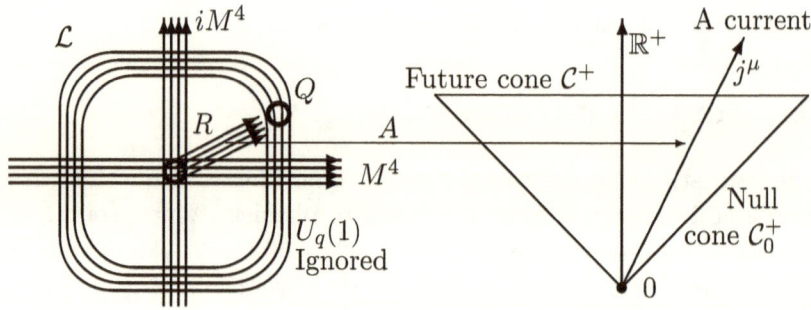

Figure 14.2: The Zovko interpretation.

The analogies between the concepts of standard quantum mechanics and their quantionic generalizations are comparatively listed in the following table.

Concepts	Complex numbers	Quantions
Real axis	$\mathbb{R}^1$	M_0^4
Imaginary axis	$i\mathbb{R}^1$	iM_0^4
Modulus	$r \in \mathbb{R}^+$	$R \in \mathcal{C}^+$
Phase factor	$e^{i\chi} \in U(1)$	$E \in U_q(1)$
Wave function	$\psi = e^{i\chi} r$	$Q = E\,R$
Algebraic norm	$\rho = A(\psi) = r^2 \in \mathbb{R}^+$	$j = A(Q) = R^2 \in \mathcal{C}^+$

14.2 Structural quantization

Observations:

(a) *Quantum mechanical concepts* are the wave function $\psi(\vec{r}, t)$, the Dirac spinor field $\Psi(x)$, and the equations of motion.

(b) *Classical concepts* are the probability density $\rho = \psi^*\psi$, the associated Schrödinger current $\vec{j} = \frac{\hbar}{i2m}(\psi^*\nabla\psi - \psi\nabla\psi^*)$, the Dirac current $j^\mu = \Psi^\dagger\gamma^0\gamma^\mu\Psi$, and the equations of continuity.

Generalization:

Definition 57 *Linear combinations of the quantionic fields $Q(x)$ and $Q^\dagger(x)$ will be referred to as **quantum** expressions. Hermitian combinations of these fields will be referred to as **classical** expressions.*

By "Hermitian combination" we refer to any sesquilinear[1] function $F = F(Q^\dagger Q)$ which satisfies the condition $F^\dagger = F$. The algebraic norm function $A(Q) = Q^\dagger Q$ is the simplest object of this type.

Zovko's interpretation yields the classical equation of continuity

$$\partial_\mu \left(Q^\dagger Q\right)^\mu = 0 \tag{14.3}$$

in the absence of an equation of motion. The latter is to be reconstructed from Equation (14.3) according to the following definitions:

Definition 58 *A **quantionic field equation** (or equation of motion) is the most general first order differential equation for $Q(x)$ whose solutions identically satisfy the equation of continuity (14.3).*

Definition 59 *The mathematical procedure that generates a quantionic field equation out of the classical equation of continuity is referred to as **structural quantization.***

Structural quantization is applied in the remaining chapters of this book to derive the Schrödinger and Dirac equations by the third procedure illustrated in Figure 14.3.

---

[1]Reminder: "Sesquilinear" means linear in Q and linear in $Q^\dagger$. One would say "bilinear" if Q and $Q^\dagger$ were not related by Hermitian conjugation.

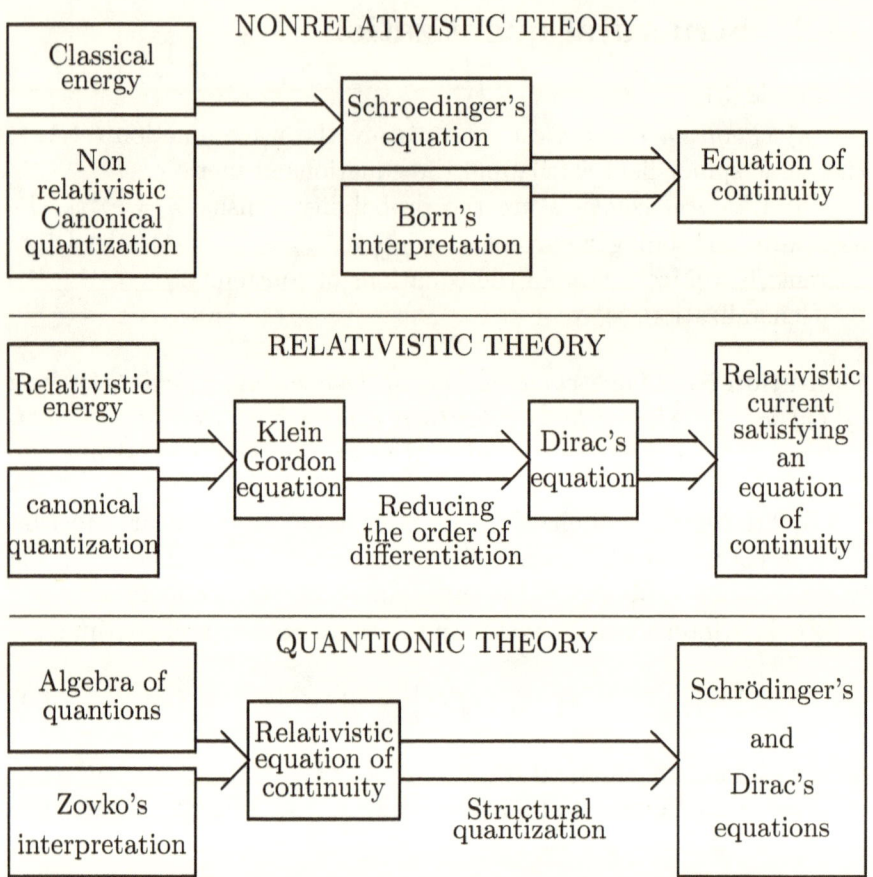

Figure 14.3: A comparison of quantum theoretic structures.

We conclude with two observations:

(1) Dirac's construction and structural quantization are analogous: The former reduces Klein-Gordon's equation of *second differential order* to Dirac's equation of *first differential order;* the latter linearizes the continuity equation of *second algebraic order* in Q to a quantionic field equation of *first algebraic order.*

(2) The equation of continuity does not suffice to determine the eight real functions that define a quantionic field $Q(x)$. We shall see that the remaining degrees of freedom are distributed among parameters of gauge transformations, components of potentials, and a mass.

Chapter 15

The Schrödinger Equation

The ideas outlined in the last chapter will now be applied to the derivation of the quantionic equation of motion in the nonrelativistic limit. This is done in several steps:

(1) The definition of the nonrelativistic limit.
(2) The application of Zovko's interpretation.
(3) The application of structural quantization.
(4) Comparison with the standard Schrödinger equation.
(5) Discussion of the results from several viewpoints.

15.1   The nonrelativistic limit

In nonrelativistic quantum mechanics, the wave function $\psi(\vec{r}, t)$ is a complex number at every spacetime point $(\vec{r}, t)$. It is thus in the center of the algebra $\mathcal{L}$ of quantions:

$$\psi(\vec{r}, t) \in \mathbb{C} \subset \mathcal{L}.$$

In the relativistic theory, whose development begins in the next chapter, the role of the complex wave function is taken over by a quantionic function.

In the transition region between the nonrelativistic quantum mechanics based on the complex wave function ψ and the relativistic quantionic theory developed in subsequent chapters, the quantionic wave function $Q = Q(\vec{r}, t)$ is in the algebra $\mathcal{L}$, but very close to the center $\mathbb{C}$ for all values of the independent variables $(\vec{r}, t)$. It may be written as a small modification of a complex function ψ,

$$Q = \psi I + Q_0, \tag{15.1}$$

where $Tr(Q_0) = 0$. (See the discussion on page 76.)

In the nonrelativistic limit, $\psi = \psi(\vec{r}, t)$ is arbitrary, while its modification $Q_0 = Q_0(\vec{r}, t)$ is assumed to be sufficiently small to be negligible at any order above the first.

To formalize the meaning of the intuitive concept of 'sufficiently small', let us write Q_0 as a linear combination of the traceless lambda matrices,

$$Q_0 = \vec{w} \cdot \Lambda, \tag{15.2}$$

where $\vec{w} = \vec{w}(\vec{r}, t)$ is a complex 3-vector. The algebraic norm of Q is then

$$
\begin{aligned}
A(Q) &= Q^\dagger Q \\
&= (\psi I + \vec{w} \cdot \Lambda)^\dagger (\psi I + \vec{w} \cdot \Lambda) \\
&= (\psi^* I + \vec{w}^{\,*} \cdot \Lambda)(\psi I + \vec{w} \cdot \Lambda).
\end{aligned}
$$

Expanded to first order in $\vec{w}$, this expression becomes

$$
\begin{aligned}
A(Q) &= \psi^* \psi + \psi^* \vec{w} \cdot \Lambda + \psi \vec{w}^{\,*} \cdot \Lambda \\
&= \psi^* \psi + (\psi^* \vec{w} + \psi \vec{w}^{\,*}) \cdot \Lambda. \tag{15.3}
\end{aligned}
$$

The discarded second-order term is

$$(\vec{w}^{\,*} \cdot \Lambda)(\vec{w} \cdot \Lambda) = (\vec{w}^{\,*} \cdot \vec{w}) I + i(\vec{w}^{\,*} \times \vec{w}) \cdot \Lambda.$$

For the future-oriented real four-vector j^μ defined by the relation

$$A(Q) = j^\mu \Lambda_\mu,$$

one obtains

$$(j^\mu) = \begin{pmatrix} j^0 \\ j^1 \\ j^2 \\ j^3 \end{pmatrix} = \begin{pmatrix} \psi^* \psi \\ w_1^* \psi + \psi^* w_1 \\ w_2^* \psi + \psi^* w_2 \\ w_3^* \psi + \psi^* w_3 \end{pmatrix}. \tag{15.4}$$

This vector consists of a scalar ρ and of a 3-vector $\vec{s}$:

$$(j^\mu) = \begin{pmatrix} \rho \\ \vec{s} \end{pmatrix} = \begin{pmatrix} \psi^*\psi \\ \psi^*\vec{w} + \psi\vec{w}^{\,*} \end{pmatrix}. \tag{15.5}$$

The time component ρ is Born's probability density. To interpret the space component $\vec{s}$ as a current $\vec{j}$, we have to make a dimensional adjustment because $\vec{s}$ and ρ have the same dimension, while a 3-current $\vec{j}$ dimensionally differs from ρ by a speed factor:

$$\vec{j} = c\vec{s}.$$

Hence, (15.5) may be rewritten as

$$\begin{pmatrix} \rho \\ \frac{1}{c}\vec{j} \end{pmatrix} = \begin{pmatrix} \psi^*\psi \\ \psi^*\vec{w} + \psi\vec{w}^{\,*} \end{pmatrix}. \tag{15.6}$$

The non-relativistic limit has the expected intuitive meaning: The vector $\vec{s}$ is infinitesimal if, at every spacetime point $(\vec{r}, t)$ where $\rho(\vec{r}, t)$ does not vanish, the ratio $\left\|\frac{\vec{j}(\vec{r},t)}{\rho(\vec{r},t)}\right\|$, which is dimensionally a speed, is sub-relativistic. In dimensionless form,

$$\left\|\frac{\psi^*\vec{w} + \psi\vec{w}^{\,*}}{\psi^*\psi}\right\| \ll 1. \tag{15.7}$$

15.2 The equation of continuity

The expression (15.6) is purely mathematical. It merely expresses the algebraic norm of a quantion in an infinitesimal neighborhood of the center $\mathbb{C}$ of the algebra $\mathcal{L}$.

Dynamics enters with the Zovko interpretation, which consists in interpreting the four-vector j^μ as a current. This means that j^μ must satisfy the equation of continuity

$$\partial_\mu j^\mu = 0.$$

In the nonrelativistic limit, the covariant derivation operator assumes the form

$$\partial_\mu = (\partial_0, \nabla),$$

so that

$$\partial_\mu j^\mu = (\partial_0, \nabla) \begin{pmatrix} \psi^*\psi \\ \psi^*\vec{w} + \psi\vec{w}\,^* \end{pmatrix}$$
$$= \partial_0 (\psi^*\psi) + \nabla \cdot (\psi^*\vec{w} + \psi\vec{w}\,^*).$$

Hence,

$$\partial_0 (\psi^*\psi) + \nabla \cdot (\psi^*\vec{w} + \psi\vec{w}\,^*) = 0 \qquad (15.8)$$

is the equation of continuity written in terms of the complex functions $\psi(\vec{r}, t)$ and $\vec{w}(\vec{r}, t)$.

15.3   Structural quantization

Given the equation of continuity (15.8), our next step is to apply the structural quantization procedure of Definition 58 on page 179. This consists in rewriting this equation in a form that allows the separation of the functions ψ^* and $\vec{w}\,^*$ from the functions ψ and $\vec{w}$. To this end, we first expand Equation (15.8):

$$\psi^* (\partial_0\psi) + (\partial_0\psi^*) \psi + (\nabla \cdot \vec{w}\,^*) \psi$$
$$+ (\nabla \cdot \vec{w}) \psi\,^* + \vec{w}\,^* \cdot (\nabla\psi) + \vec{w} \cdot (\nabla\psi\,^*)$$
$$= 0.$$

This expression admits two different ways of regrouping the six terms into three:

$$\mathrm{Re}\,[\psi^* (\partial_0\psi + \nabla \cdot \vec{w}) + \vec{w}\,^* \cdot (\nabla\psi)] = 0,$$

and

$$\mathrm{Re}\,[\psi^* (\partial_0\psi + \nabla \cdot \vec{w}) + \vec{w} \cdot (\nabla\psi\,^*)] = 0.$$

These equations are simultaneously satisfied if and only if

$$\mathrm{Re}\,[\psi^* (\partial_0\psi + \nabla \cdot \vec{w})] = 0,$$
$$\mathrm{Re}\,[\vec{w}\,^* \cdot (\nabla\psi)] = 0.$$

The most general solutions are

$$\partial_0\psi + \nabla \cdot \vec{w} = iF\psi, \qquad (15.9)$$
$$\nabla\psi = iN\vec{w}, \qquad (15.10)$$

where $F = F(\vec{r}, t)$ is an arbitrary real function, and, in the most general case, $N = N(\vec{r}, t)$ is a real 3-tensor.

Thus, after structural quantization in the nonrelativistic limit, the single non-linear classical equation of continuity (15.8) is replaced by a system of linear differential equations for the quantum mechanical functions ψ and $\vec{w}$. One can collect the coupled equations (15.9) and (15.10) into a single differential equation in matrix form:

$$\begin{pmatrix} \partial_0 & \partial_1 & \partial_2 & \partial_3 \\ \partial_1 & 0 & 0 & 0 \\ \partial_2 & 0 & 0 & 0 \\ \partial_3 & 0 & 0 & 0 \end{pmatrix} \begin{pmatrix} \psi \\ w_1 \\ w_2 \\ w_3 \end{pmatrix} = i \begin{pmatrix} F & 0 & 0 & 0 \\ 0 & N_{11} & N_{12} & N_{13} \\ 0 & N_{21} & N_{22} & N_{23} \\ 0 & N_{31} & N_{32} & N_{33} \end{pmatrix} \begin{pmatrix} \psi \\ w_1 \\ w_2 \\ w_3 \end{pmatrix}.$$

This completes the structural quantization of the nonrelativistic equation of continuity. In the next step, we shall derive the standard form of the Schrödinger equation.

15.4 Schrödinger's equation

We shall decouple the-first order differential equations (15.9) and (15.10) by going to second order. To this end, we assume that the tensor N has an inverse. Hence, dividing Equation (15.10) by N yields

$$\vec{w} = -iN^{-1}\nabla\psi. \tag{15.11}$$

By substituting this expression and

$$\partial_0 = \frac{1}{c}\frac{\partial}{\partial t} \tag{15.12}$$

into Equation (15.9), one obtains

$$\frac{1}{c}\frac{\partial}{\partial t}\psi + \nabla \cdot \left(-iN^{-1}\nabla\psi\right) = iF\psi.$$

Expansion of the gradient yields

$$\frac{1}{c}\frac{\partial}{\partial t}\psi - iN^{-1}\nabla^2\psi - i\left(\nabla \cdot N^{-1}\right) \cdot (\nabla\psi) = iF\psi. \tag{15.13}$$

This equation generalizes Schrödinger's equation because it contains all the terms of the latter:

(a) There is a Laplacian term: $\nabla^2\psi$.

(b) There is a first time derivative: $\frac{\partial}{\partial t}\psi$.

(c) There is a linear term: $F\psi$.

(d) The term $\left(\nabla \cdot N^{-1}\right) \cdot \left(\nabla\psi\right)$ is additional, but it vanishes if N is a constant matrix.

Equation (15.13) is thus the generalized Schrödinger equation for the case of a position-dependent mass tensor N.

Since all terms of the standard Schrödinger equation

$$\left(-\frac{\hbar^2}{2m}\nabla^2 + V\right)\psi = i\hbar\frac{\partial}{\partial t}\psi \tag{15.14}$$

are present in Equation (15.13), one obtains the interpretations of the coefficients of the latter by comparing these two equations. To this end, we multiply all terms of Equation (15.13) by $ic\hbar$, and then rearrange the resulting equation so that its right-hand side would coincide with the right-hand side of Equation (15.14):

$$-c\hbar N^{-1}\nabla^2\psi - c\hbar\left(\nabla N^{-1}\right) \cdot \left(\nabla\psi\right) - c\hbar F\psi = i\hbar\frac{\partial}{\partial t}\psi.$$

Term-by-term comparison of this equation with the Schrödinger Equation (15.14) yields

$$V = -c\hbar F,$$
$$-\frac{\hbar^2}{2m} = -c\hbar N^{-1}.$$

It follows that the function F is essentially the potential,

$$F\left(\vec{r}, t\right) = -\frac{1}{c\hbar}V\left(\vec{r}, t\right), \tag{15.15}$$

and that N is not a tensor but a scalar:

$$N = \frac{2cm}{\hbar}. \tag{15.16}$$

The mass parameter m may nevertheless be a function of space, in which case substitution of (15.16) into Equation (15.13) yields the generalized Schrödinger equation

$$-\frac{\hbar^2}{2m}\nabla^2\psi - \frac{\hbar^2}{2}\left(\nabla\frac{1}{m}\right) \cdot \left(\nabla\psi\right) + V\psi = i\hbar\frac{\partial}{\partial t}\psi, \tag{15.17}$$

which can be rewritten in the more elegant compact form

$$\left(-\frac{\hbar^2}{2}\nabla\frac{1}{m}\nabla + V\right)\psi = i\hbar\frac{\partial}{\partial t}\psi. \tag{15.18}$$

Let us point out that the concept of a space-dependent mass has been used in solid state physics, where the freedom of choice of the mass function $m = m(\vec{r})$ is exploited to model the effects of the lattice on the propagation of the wave function. The first task in this approach was to correctly select the Hamiltonian supposed to generalize the Schrödinger operator. Since the only condition available to fix the Hamiltonian is Hermiticity, the solution is not unique. Hence, two different generalizations of the operator $\frac{1}{m}\nabla^2$ have been considered in the literature:

1st choice: $\qquad \dfrac{1}{m}\nabla^2 \rightarrow \nabla\dfrac{1}{m}\nabla,$

2nd choice: $\qquad \dfrac{1}{m}\nabla^2 \rightarrow \dfrac{1}{2}\left(\dfrac{1}{m}\nabla^2 + \nabla^2\dfrac{1}{m}\right).$

We see that the first choice is the only one consistent with the quantionic version of Schrödinger's equation.

15.5 Discussions

The Schrödinger equation having been derived as a theorem in the quantionic approach, the purpose of the following subsections is to verify the well known properties of this equation within the same approach.

Schrödinger's current

A 3-current $\vec{j}$ is given *a priori* by Zovko's interpretation. It is the spatial part of the expression (15.6), that is

$$\vec{j} = c\left(\psi^*\vec{w} + \psi\vec{w}^{\,*}\right). \tag{15.19}$$

Substitution of the expression for N given by relation (15.16) into the expression for $\vec{w}$ given by relation (15.11) yields

$$\vec{w} = \frac{\hbar}{i2cm}\nabla\psi. \tag{15.20}$$

Hence, the expression (15.19) for the current of quantionic origin assumes the form

$$\vec{j} = c\left(\psi^* \frac{\hbar}{i2cm}\nabla\psi + \psi\left(\frac{\hbar}{i2cm}\nabla\psi\right)^*\right)$$

$$= \frac{\hbar}{i2m}\left(\psi^*\nabla\psi - \psi\nabla\psi^*\right). \qquad (15.21)$$

We recognize in this result the standard textbook expression for the current derived from the Schrödinger Equation (15.13) and the equation of continuity

$$\frac{\partial}{\partial t}\left(\psi^*\psi\right) + \nabla\cdot\vec{j} = 0. \qquad (15.22)$$

It is instructive to re-derive Schrödinger's current by the standard textbook procedure, but in the presence of the additional term which stems from a position-dependent mass. To this end, we multiply Equation (15.17) from the left by ψ^*,

$$-\frac{\hbar^2}{2m}\psi^*\nabla^2\psi - \frac{\hbar^2}{2}\left(\nabla\frac{1}{m}\right)\cdot\psi^*\left(\nabla\psi\right) + V\psi^*\psi = i\hbar\psi^*\frac{\partial}{\partial t}\psi,$$

and we similarly multiply its complex conjugate by ψ,

$$-\frac{\hbar^2}{2m}\psi\nabla^2\psi^* - \frac{\hbar^2}{2}\left(\nabla\frac{1}{m}\right)\cdot\psi\left(\nabla\psi^*\right) + \psi V\psi^* = -i\hbar\psi\frac{\partial}{\partial t}\psi^*.$$

By subtracting the first equation from the second, one obtains

$$\frac{\hbar^2}{2m}\left(\psi^*\nabla^2\psi - \psi\nabla^2\psi^*\right) + \frac{\hbar^2}{2}\left(\nabla\frac{1}{m}\right)\cdot\left(\psi^*\nabla\psi - \psi\nabla\psi^*\right)$$

$$= -i\hbar\frac{\partial}{\partial t}\left(\psi^*\psi\right). \qquad (15.23)$$

The second term on the left-hand side may be restructured as follows:

$$\frac{\hbar^2}{2}\left(\nabla\frac{1}{m}\right)\cdot\left(\psi^*\nabla\psi - \psi\nabla\psi^*\right)$$

$$= \frac{\hbar^2}{2}\nabla\cdot\left(\frac{1}{m}\left(\psi^*\nabla\psi - \psi\nabla\psi^*\right)\right) - \frac{\hbar^2}{2m}\nabla\cdot\left(\psi^*\nabla\psi - \psi\nabla\psi^*\right)$$

$$= \frac{\hbar^2}{2}\nabla\cdot\left(\frac{1}{m}\left(\psi^*\nabla\psi - \psi\nabla\psi^*\right)\right) - \frac{\hbar^2}{2m}\left(\psi^*\nabla^2\psi - \psi\nabla^2\psi^*\right).$$

Substitution of this expression into Equation (15.23) yields

$$\nabla \cdot \frac{\hbar}{2im} \left(\psi^* \nabla \psi - \psi \nabla \psi^* \right) + \frac{\partial}{\partial t} \left(\psi^* \psi \right) = 0, \tag{15.24}$$

which is the equation of continuity for the probability density $\rho = \psi^* \psi$ and the Schrödinger current (15.21) in which the mass is a function of spacial coordinates.

Thus, the formal expression for the Schrödinger current $\vec{j}$ is the same, whether derived from Schrödinger's equation or extracted from Zovko's interpretation. In the latter case, is not affected by the space-dependence of the mass.

While many algebraic and geometric results derived in Part I support the physicalness of the quantionic approach, they are all static. The quantionic derivation of the Schrödinger equation is the first dynamical result which confirms the correctness of Zovko's interpretation and of the structural quantization procedure that follows it.

The range of validity

The quantum theory obtained from classical mechanics by applying the canonical quantization rules (14.1) does not impose any upper limit on the allowed values of speeds. We know that the theory is nonrelativistic because is stems from classical mechanics, but this is not an internally generated restriction.

In contrast, the quantionic derivation of nonrelativistic quantum mechanics by the structural quantization of Zovko's interpretation of the inherently causal vector $j = A(Q)$ does provide a numerical measure for the domain of validity of this theory.

To see how, let us begin with several observations concerning the quantionic derivation of the Schrödinger equation in Section 15.4:

Clearly, this equation cannot be derived within the field $\mathbb{C}$ of complex numbers because the algebraic norm $A(\psi)$ is a scalar — not a vector that can be interpreted as a current. Similarly, it cannot be derived very far from the field $\mathbb{C}$ of complex numbers, where the square of the traceless part of Q in the current $j = A(Q)$ is not sufficiently small to be negligible.

Referring to the diagram on page 182, the derivation of the Schrödinger equation takes place in a narrow region of the algebra $\mathcal{L}$ of quantions which is adjacent to the algebra's center $\mathbb{C}$. It is in this infinitesimal transition region that separate relativistic from nonrelativistic physics that Zovko's interpretation gives rise to a nonrelativistic equation of continuity.

The algebraic condition that the quantionic function $Q(\vec{r}, t)$ is to remain in a small neighborhood of the complex wave function $\psi(\vec{r}, t)$ for all values of $\vec{r}$ and all values of t is equivalent to the geometric condition that the dimensionless current

$$j^\mu = \begin{pmatrix} \rho \\ \vec{s} \end{pmatrix},$$

given by Equation (15.5), is to remain in a small neighborhood of the axis $\mathbb{R}^+$ of the future-oriented cone C^+. This condition is illustrated in Figure 15.1, where the 'sub-relativistic cone' is shown very large for convenience.

Note that working with the 3-vector $\vec{s}$ instead of with the physical current $\vec{j}$ is equivalent to taking $c = 1$, which is what we need to derive dimensionless conditions.

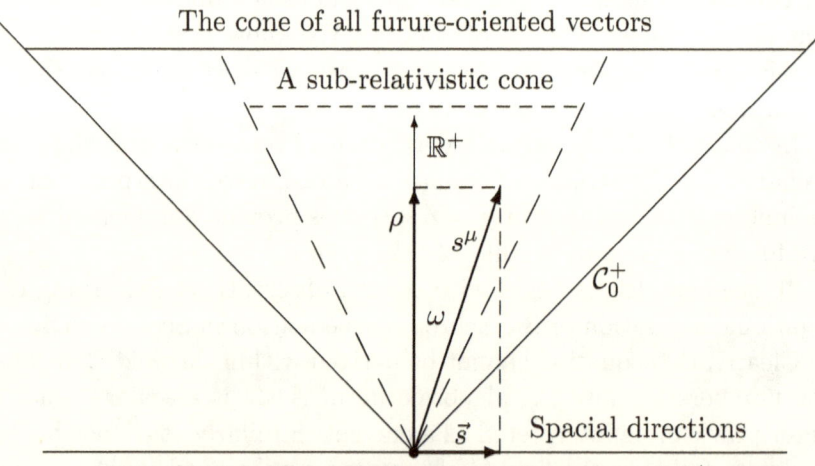

Figure 15.1: The neighborhood of the axis of C^+.

In terms the geometric variables of Figure 15.1, the condition (15.7) which defines the nonrelativistic limit is equivalent to the dimensionless condition

$$\tanh \omega = \frac{\|\vec{s}\|}{\rho} \ll 1. \tag{15.25}$$

To compute the nominator $\|\vec{s}\|$, we take the expression for the vector $\vec{s} = \frac{1}{c}\vec{j}$ from relation (15.21),

$$\vec{s} = \frac{\hbar}{i2cm} \left(\psi^* \nabla \psi - \psi \nabla \psi^* \right).$$

By the triangle inequality, the length of this vector satisfies the following condition:

$$\begin{aligned}
\|\vec{s}\| &= \frac{\hbar}{2cm} \|\psi^* \nabla \psi - \psi \nabla \psi^*\| < \frac{\hbar}{cm} \|\psi^* \nabla \psi\| \\
&= \frac{\hbar}{cm} \|\nabla \psi\| \cdot \|\psi^*\| = \frac{\hbar}{cm} \|\nabla \psi\| \cdot \|\psi\| .
\end{aligned} \tag{15.26}$$

The substitution of the expression

$$\rho = \psi^* \psi = \|\psi\|^2$$

for ρ and of the inequality (15.26) for $\|\vec{s}\|$ into relation (15.25) yields the condition

$$\frac{\hbar}{cm} \frac{\|\nabla \psi\|}{\|\psi\|} \ll 1.$$

Since the coefficient $\frac{\hbar}{cm}$ is the Compton wavelength of the particle described by the wave function ψ,

$$\lambda_m = \frac{\hbar}{cm}, \tag{15.27}$$

the nonrelativistic limit is defined by the simple inequality

$$\left\| \frac{\nabla \psi}{\psi} \right\| \ll \frac{1}{\lambda_m}. \tag{15.28}$$

In words:

Condition 60 *For the Schrödinger equation of a particle to be valid, the logarithmic gradient $\nabla \ln \psi = \frac{\nabla \psi}{\psi}$ of the wave function must, at all spacetime points $(\vec{r}, t)$, be negligible in comparison with the reciprocal Compton wavelength of the particle.*

As usual in physics, the meaning of 'negligible' is defined by the accuracy of the measurements.

The condition (15.28) implies that the more variable a wave function is in a region of space, the more likely it is to be nonrelativistic. This might impose conditions on the allowed potentials.

Similarly, in the case of a plane wave, the shortest wavelengths are most relativistic. To verify this last statement, let us consider a plane wave

$$\psi = e^{i(kx - \omega t)}$$

in a region of space where $V = 0$. The condition (15.28) reads

$$\lambda_m \left\| \frac{\nabla \psi}{\psi} \right\| = \lambda_m \left\| ik \right\| = \lambda_m k \ll 1.$$

The de Broglie identification

$$k = \frac{p}{\hbar}$$

implies

$$k = \frac{mv}{\hbar} = \frac{mc}{\hbar} \beta = \frac{1}{\lambda_m} \beta,$$

where $\beta = v/c$. Substitution of this expression into the previous inequality yields

$$\beta \ll 1,$$

which is the standard kinematic definition of the nonrelativistic region. This result is consistent with our initial definition of the nonrelativistic limit.

Comparison of the two approaches

Historically, the Schrödinger equation was postulated to yield spectra of eigenvalues that matched experimental spectroscopic data. We may also regard it as derived from classical mechanics by canonical quantization, in which case it is the canonical quantization rules that are postulated. In either case, let us refer to the quantum theory based on postulates — which is the standard quantum theory built over the field of complex numbers — as the "postulated quantum mechanics".

In the quantionic approach, by contrast, the Schrödinger equation is a theorem that stems from a single postulate, which is Zovko's interpretation of the algebraic norm, this norm being an inherently causal Minkowski vector. As for the structural quantization that follows, it is not a postulate but a linearization procedure based on a separation of variables. To distinguish it from the postulated quantum mechanics, we shall refer to the quantum theory of quantionic origin as the "derived quantum mechanics".

It is instructive to compare the two approaches graphically — especially since analogous observations will be applicable to the difference between the "postulated Dirac equation" and the "derived Dirac equation". The comparison is encapsulated in the self-explanatory Figure 15.2, where 'BEFORE' and 'AFTER' refer to 'before interpretation' and 'after interpretation'.

To follow this diagram, we begin with the mathematical theory of quantions and its physical interpretations. At this point, the latter consist only of Zovko's interpretation. Thus, the top box is relativistic.

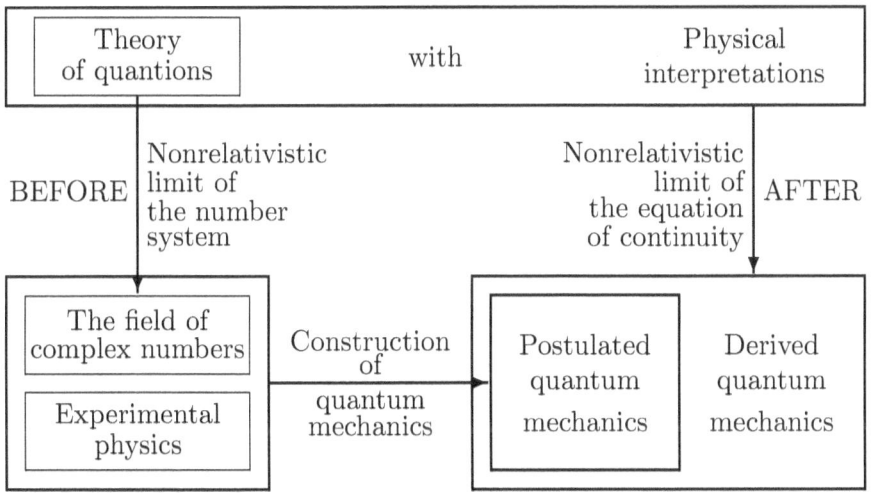

Figure 15.2: Two limiting procedures.

The nonrelativistic limit may be taken either before or after the physical interpretation of quantions.

If taken before the interpretation, as shown on the left-hand side of the diagram, the limit is the field of complex numbers. Thus, going to the nonrelativistic limit at this point leaves the dynamics behind, that is, it does not take along the differential equation of continuity. Consequently, external insights from experimental physics are needed to construct the Schrödinger equation in the "postulated quantum mechanics".

If the limit is taken after the interpretation, as shown on the right-hand side of the diagram, it carries with it all the physics needed to yield the "derived quantum mechanics" as an exact theorem. The derived quantum mechanics includes the postulated quantum mechanics as well as additional insights — or additional physics.

In the nonrelativistic limit, the main additional insights are that it is theoretically legitimate to regard the mass as a nowhere vanishing scalar field in 3-space that, and that

$$H = \frac{\hbar^2}{2} \nabla \frac{1}{m} \nabla + V$$

is the form of the most general space-dependent Schrödinger Hamiltonian.

In the fully relativistic version, the additional physics identified so far (in Part II) is the electroweak gauge group $SU(2) \times U(1)$. In contrast, this group is only experimentally justified in the Standard Model.

Chapter 16

The Quantionic Scalar Field Equation

In the nonrelativistic limit, the derivation of the equation of motion (the Schrödinger equation) and its complete discussion were simple enough to be completed in a single chapter.

In the relativistic case, the derivation of the equation of motion (the quantionic field equation) is also simple enough to be derived on a few pages in the present chapter, but this equation is structurally so rich that the remaining chapters of the book are dedicated to its discussion. Even so, much more is left to be investigated and understood in terms of Standard Model concepts.

After a conceptual introduction to the subject in Section 16.1, the relativistic quantionic field equation is derived in Section 16.2.

16.1 An overview of results

In classical physics, real functions of space and time are referred to as (real) scalar fields. In quantum physics, complex functions in space-time are referred to as (complex) scalar fields. Generalizing, a scalar field is a single-component field, the component being an element of the number system of the theory in question.

Examples of multi-component fields are tensors with real or complex components, and vectors in Hilbert space.

We extend this terminology to the new number system of quantions. A single-component quantionic function $Q(x) \in \mathcal{L}$ is thus a "quantionic scalar field" in spacetime. The latter is limited to the affine Minkowski space in the present work.

Just as complex fields are viewed as single objects, not as pairs of mutually independent real fields, preservation of structure requires that quantionic scalar fields be viewed as single objects — either as matrices $Q(x)$ or vectors $|q(x)\rangle$ — not as quadruples of mutually independent complex fields. The proofs of theorems may be performed with the components q_i treated as independent variables, but the theorems themselves must be formulated in structural terms.

As shown in the present chapter, Zovko's interpretation followed by structural quantization yields a very general relativistic quantionic scalar field equation. The procedure that leads to it, and the subsequent interpretations of the related objects as potentials and mass, consist of the following four steps.

First step (the present chapter) Reminder: In the nonrelativistic case of Chapter 15, structural quantization yields a single complex differential equation which contains two arbitrary separating functions: a potential function and a mass function. Both are real functions of space and time.

Preview of the result obtained in this chapter: In the relativistic case, structural quantization yields a single quantionic differential equation which contains two arbitrary separating functions: a potential function and a mass function. The potential is represented by a Hermitian 4×4 matrix (16 real functions), and the mass by an complex antisymmetric 4×4 matrix (12 real functions).

A difference between the nonrelativistic Schrödinger equation and the relativistic quantionic differential equation is that former is of second order while the latter is of the first order. There is nothing profound about this, as one can always trade orders of differentiation against degrees of freedom. Thus, a first order matrix version of the nonrelativistic equation is displayed on page 185.

Let us also point out that the quantionic field $Q(x) \in \mathcal{L}$ appears in differential equations in its linear representation $|q(x)\rangle \in \mathcal{H}_q$. This is consistent with the conclusions of Chapter 13.

Second step (Chapter 17)   In the general quantionic field equation, the mass matrix couples the field $|q(x))$ to its complex conjugate $|q^*(x))$. Stated differently, the field $|q^*)$ appears as a source in the linear differential equation for the field $|q)$. The opposite is true in the complex conjugate equation. It follows that the fields $|q)$ and $|q^*)$ are 'doubly coupled': They are coupled differentially by the field equation, and algebraically by complex conjugation — which means that they are overdetermined. Therefore, there exists no solution unless some consistency conditions are met. These conditions are identified in Chapter 17, where they reduce the number of real parameters in the mass matrix from twelve to four.

Third step (Chapter *18*)   The structuring of the algebra of quantions into L-type and R-type matrices induces a decomposition of the 16-parametric Hermitian matrix potential into four vector potentials. One of these is the standard electromagnetic potential. The other three are natural candidates for the intermediate vector boson fields, but the exact physical interpretations of these potentials are beyond the scope of the present volume.

Fourth step (Chapter *19*)   The physicalness of the quantionic approach is confirmed in this chapter. This is done by establishing a one-to-one correspondence between quantionic scalars and Dirac spinors, and by proving that the general quantionic field equation implies Dirac's equation. Yet, the former contains more physics.

The number system

— In quantum field theory, the number system is the field of complex numbers. This system has been taken over from nonrelativistic quantum mechanics owing to an apparent absence of a reason to do otherwise and to the actual absence of a known alternative.

— In the quantionic approach, the number system is the algebra of quantions. This algebra is postulated in the present work after having been heuristically derived elsewhere.

Relativity

— In quantum field theory, the relativistic field equations are derived from relativistic Lagrangians hat are selected so as to yield field equations that agree with observations.

— In the quantionic approach, the field equations follow from the structure of the number system augmented only with Zovko's interpretation. The procedure is the same as in the derivation of Schrödinger's equation, but it is exact — that is, there are no approximations.

Gauge groups

— In quantum field theory, the gauge group $SU(3) \times SU(2) \times U(1)$ is selected so as to restrict the choice of Lagrangians to those which yield field equations in agreement with observations.

— In the quantionic approach, which has been developed so far only for quantionic scalar fields, the electroweak gauge group $SU(2) \times U(1)$ is a fundamental algebraic property of quantions.

Spinors

— In quantum field theory, the Dirac spinors appear as a representation of the Lorentz group.

— In the quantionic approach, the Dirac spinors are simply quantions — but 'viewed from a different angle'. They are defined in an unguessable way as mixtures of the complex quantionic components q_i and q_j^*.

The Dirac operator

— In quantum field theory, the fundamental differential operator $D = \gamma^\mu \partial_\mu$ is the algebraic square root of the D'Alambertian:

$$\Box = D^\dagger D \equiv D^2.$$

— In the quantionic approach, the fundamental differential operator $\mathcal{D}$ is the metric square root of the D'Alambertian:

$$\Box = \mathcal{D}^\# \mathcal{D}.$$

16.2 Derivation of the equation of motion

Quantionic scalar fields will be interchangeably represented, as needed, by three kinds of objects:

(1) By L-type matrices when the algebraic product of quantions is relevant.

(2) By Minkowski vectors in the quantionic Gaussian space when classical differential equations are relevant.

(3) By q-kets in quantionic field equations.

The relationships between these three formally different representation of the same mathematical object will be considered known for having been developed in Part I.

A straightforward procedure that leads to the equation of motion, that is, to the general quantionic scalar field equation, consists of exactly three steps. Already discussed on several occasions, these steps are charted for convenience in Figure 16.1.

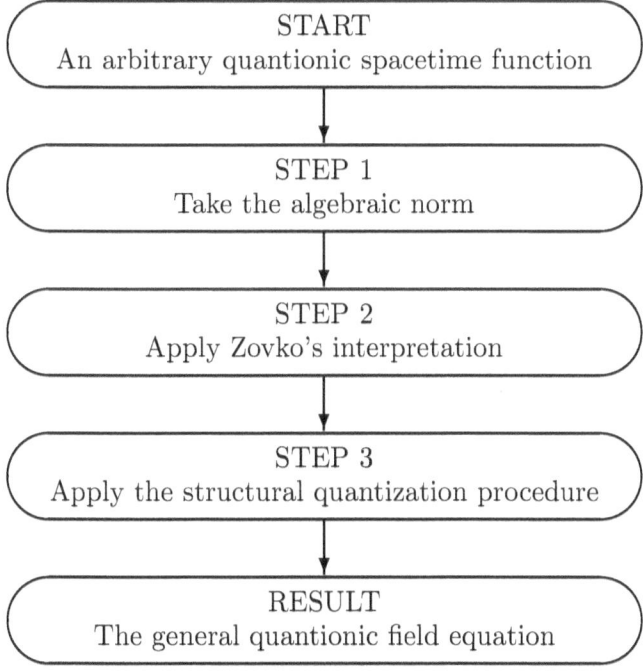

Figure 16.1: The derivation of the field equation.

It is interesting to note that this general procedure, which yields a relativistic quantionic field equation, is simpler than the procedure that gave us the Schrödinger equation. The reason is that the latter needs an additional step: The transition to the nonrelativistic limit.

In the following application of the charted procedure, a subsection is dedicated to each step.

Step 1: The algebraic norm

The calculations will be done explicitly in the matrix formalism. We begin with an L-type quantionic function

$$Q\left(x\right) = \begin{pmatrix} q_1 & q_3 & 0 & 0 \\ q_2 & q_4 & 0 & 0 \\ 0 & 0 & q_1 & q_3 \\ 0 & 0 & q_2 & q_4 \end{pmatrix}. \tag{16.1}$$

Its Hermitian conjugate is

$$Q^\dagger\left(x\right) = \begin{pmatrix} q_1^* & q_2^* & 0 & 0 \\ q_3^* & q_4^* & 0 & 0 \\ 0 & 0 & q_1^* & q_2^* \\ 0 & 0 & q_3^* & q_4^* \end{pmatrix}. \tag{16.2}$$

Its algebraic norm is

$$A\left(Q\right) = Q^\dagger\left(x\right) Q\left(x\right) =$$
$$\begin{pmatrix} q_1^* q_1 + q_2^* q_2 & q_1^* q_3 + q_2^* q_4 & 0 & 0 \\ q_3^* q_1 + q_4^* q_2 & q_3^* q_3 + q_4^* q_4 & 0 & 0 \\ 0 & 0 & q_1^* q_1 + q_2^* q_2 & q_1^* q_3 + q_2^* q_4 \\ 0 & 0 & q_3^* q_1 + q_4^* q_2 & q_3^* q_3 + q_4^* q_4 \end{pmatrix} \tag{16.3}$$

To interpret this norm as a four-current $j^\mu\left(x\right)$, we use the mapping (3) in Figure 11.3 on page 134, relation (11.18),

$$A\left(Q\right) \rightleftarrows j^\mu = \frac{1}{4} Tr\left[A\left(Q\right)\Lambda_\mu\right].$$

In components:

$$\left(j^\mu\right) \equiv \begin{pmatrix} j^0 \\ j^1 \\ j^2 \\ j^3 \end{pmatrix} = \frac{1}{2}\begin{pmatrix} q_1^* q_1 + q_2^* q_2 + q_3^* q_3 + q_4^* q_4 \\ q_3^* q_1 + q_4^* q_2 + q_1^* q_3 + q_2^* q_4 \\ i\left(q_1^* q_3 + q_2^* q_4 - q_3^* q_1 - q_4^* q_2\right) \\ q_1^* q_1 + q_2^* q_2 - q_3^* q_3 - q_4^* q_4 \end{pmatrix}. \tag{16.4}$$

Step 2: The equation of continuity

The calculations in this section are presented in detail in a 'pedestrian' way, and independently of the work done in Chapter 13.

The Zovko interpretation implies that the four-vector $j(x)$ must satisfy the equation of continuity

$$(\partial, j) \overset{def}{=} \partial_\mu j^\mu = 0. \tag{16.5}$$

Written out in full, the divergence is

$$\partial_\mu j^\mu = \partial_0 j^0 + \partial_1 j^1 + \partial_2 j^2 + \partial_3 j^3. \tag{16.6}$$

After some consolidations, substitution of the expressions (16.4) into Equation (16.6) yields the following relation:

$$\begin{aligned}
2\partial_\mu j^\mu = \; & q_1^* \left[(\partial_0 + \partial_3) q_1 + (\partial_1 + i\partial_2) q_3 \right] \\
& + q_2^* \left[(\partial_0 + \partial_3) q_2 + (\partial_1 + i\partial_2) q_4 \right] \\
& + q_3^* \left[(\partial_0 - \partial_3) q_3 + (\partial_1 - i\partial_2) q_1 \right] \\
& + q_4^* \left[(\partial_0 - \partial_3) q_4 + (\partial_1 - i\partial_2) q_2 \right] \\
& + \left[(\partial_0 + \partial_3) q_1^* + (\partial_1 - i\partial_2) q_3^* \right] q_1 \\
& + \left[(\partial_0 + \partial_3) q_2^* + (\partial_1 - i\partial_2) q_4^* \right] q_2 \\
& + \left[(\partial_0 - \partial_3) q_3^* + (\partial_1 + i\partial_2) q_1^* \right] q_3 \\
& + \left[(\partial_0 - \partial_3) q_4^* + (\partial_1 + i\partial_2) q_2^* \right] q_4.
\end{aligned}$$

This expression can be simplified by assigning special symbols to the pairs of partial derivation operators in parentheses. These have already been introduced as the Newman-Penrose symbols:

$$\left. \begin{aligned}
D &\overset{def}{=} \partial_0 + \partial_3, \\
\Delta &\overset{def}{=} \partial_0 - \partial_3, \\
\delta &\overset{def}{=} \partial_1 + i\partial_2, \\
\delta^* &\overset{def}{=} \partial_1 - i\partial_2.
\end{aligned} \right\} \tag{16.7}$$

In terms of these derivatives, the equation of continuity assumes the more compact form

$$
\begin{aligned}
0 = {}& q_1^* \left[Dq_1 + \delta q_3 \right] + \left[Dq_1^* + \delta^* q_3^* \right] q_1 \\
& + q_2^* \left[Dq_2 + \delta q_4 \right] + \left[Dq_2^* + \delta^* q_4^* \right] q_2 \\
& + q_3^* \left[\Delta q_3 + \delta^* q_1 \right] + \left[\Delta q_3^* + \delta q_1^* \right] q_3 \\
& + q_4^* \left[\Delta q_4 + \delta^* q_2 \right] + \left[\Delta q_4^* + \delta q_2^* \right] q_4 .
\end{aligned}
$$

The expressions in square brackets can be collected into a q-ket. Since this vector is constructed from various Newman-Penrose derivatives of the components of q, it is suggestive to denote it by $|\mathcal{D}q)$, where $\mathcal{D}$ is a differential operator that remains to be determined.

$$
|\mathcal{D}q) \stackrel{def}{=} \begin{pmatrix} Dq_1 + \delta q_3 \\ Dq_2 + \delta q_4 \\ \Delta q_3 + \delta^* q_1 \\ \Delta q_4 + \delta^* q_2 \end{pmatrix} . \tag{16.8}
$$

The Hermitian conjugate of the q-ket $|\mathcal{D}q)$ is the q-bra $(\mathcal{D}q|$.

The divergence (∂, j) now assumes the compact form

$$
(\partial, j) = \frac{1}{2} \left[(q|\mathcal{D}q) + (\mathcal{D}q|q) \right] . \tag{16.9}
$$

An even shorter and more useful expression is

$$
(\partial, j) = \operatorname{Re} (q|\mathcal{D}q) . \tag{16.10}
$$

The components of the q-ket (16.8) are such that the 'symbolic' differential operator $\mathcal{D}$ can be extracted out of $|\mathcal{D}q)$ as a 'concrete' differential matrix operator defined by the identity

$$
\mathcal{D} |q) \equiv |\mathcal{D}q) . \tag{16.11}
$$

Thus:

$$
\mathcal{D} |q) \equiv \begin{pmatrix} D & 0 & \delta & 0 \\ 0 & D & 0 & \delta \\ \delta^* & 0 & \Delta & 0 \\ 0 & \delta^* & 0 & \Delta \end{pmatrix} \begin{pmatrix} q_1 \\ q_2 \\ q_3 \\ q_4 \end{pmatrix} = \begin{pmatrix} Dq_1 + \delta q_3 \\ Dq_2 + \delta q_4 \\ \Delta q_3 + \delta^* q_1 \\ \Delta q_4 + \delta^* q_2 \end{pmatrix} . \tag{16.12}
$$

As a matrix, the operator $\mathcal{D}$ is evidently Hermitian — in the sense that if $f(x)$ is a real function, then $\mathcal{D}f$ is a Hermitian matrix.

Definition 61 *The Hermitian differential matrix operator*

$$\mathcal{D} \overset{def}{=} \begin{pmatrix} D & 0 & \delta & 0 \\ 0 & D & 0 & \delta \\ \delta^* & 0 & \Delta & 0 \\ 0 & \delta^* & 0 & \Delta \end{pmatrix} \tag{16.13}$$

will be referred to as the **quantionic derivation operator.**

Reminder: A *differential operator* is any operator that contains partial derivatives. A *derivation operator* is a differential operator that satisfies the Leibniz identity.

Formally, $\mathcal{D} \in \mathcal{R}$, and its elements are linear combinations of the partial derivatives ∂_μ. Hence, by an argument developed in Chapter 13, $\mathcal{D}$ is a derivation operator. We also observe that $\mathcal{D}$ is identical to $\mathcal{D}_1$ defined by (13.42). To avoid mixing the exact results of this chapter with those of Chapter 13, which are based in part on heuristics, we shall no longer refer to that chapter.

Theorem 62 *The metric norm of the quantionic derivation operator is identical to the metric norm of the gradient operator ∂_μ :*

$$m(\mathcal{D}) = \eta^{\mu\nu} \partial_\mu \partial_\nu \equiv \Box. \tag{16.14}$$

Proof. By the formal definition of the metric norm

$$M(\mathcal{D}) = \mathcal{D}^\# \mathcal{D}.$$

The metric dual, or parity transform, of the matrix operator $\mathcal{D}$ is

$$\mathcal{PD} = \mathcal{D}^\# = \begin{pmatrix} \Delta & 0 & -\delta & 0 \\ 0 & \Delta & 0 & -\delta \\ -\delta^* & 0 & D & 0 \\ 0 & -\delta^* & 0 & D \end{pmatrix}. \tag{16.15}$$

Hence,

$$M(\mathcal{D}) = \begin{pmatrix} \Delta & 0 & -\delta & 0 \\ 0 & \Delta & 0 & -\delta \\ -\delta^* & 0 & D & 0 \\ 0 & -\delta^* & 0 & D \end{pmatrix} \begin{pmatrix} D & 0 & \delta & 0 \\ 0 & D & 0 & \delta \\ \delta^* & 0 & \Delta & 0 \\ 0 & \delta^* & 0 & \Delta \end{pmatrix}$$
$$\equiv m(\mathcal{D}) \, I = (\Delta D - \delta \delta^* (\mathcal{D})) \, I. \tag{16.16}$$

Substitution of the expressions (16.7) into this solution for the metric norm $m(\mathcal{D})$ yields

$$
\begin{aligned}
m(\mathcal{D}) &= D\Delta - \delta\delta^* \\
&= (\partial_0 + \partial_3)(\partial_0 - \partial_3) - (\partial_1 + i\partial_2)(\partial_1 - i\partial_2) \\
&= \partial_0^2 - \partial_1^2 - \partial_2^2 - \partial_3^2 = \eta^{\mu\nu}\partial_\mu\partial_\nu = \square, \qquad (16.17)
\end{aligned}
$$

proving the assertion. ∎

Using the expression (16.10) for the divergence and the subsequent definition of the quantionic derivation operator, the equation of continuity simplifies to

$$
\mathrm{Re}\,(q|\,\mathcal{D}\,|q) = 0. \qquad (16.18)
$$

This is the formal expression of Zovko's interpretation stated in terms of quantions. An equivalent statement of condition (16.18) is that the expression $(q|\,\mathcal{D}\,|q)$, which is a complex number in principle, should be purely imaginary:

$$
(q|\,\mathcal{D}\,|q) = \text{Imaginary}. \qquad (16.19)
$$

Step 3: Structural quantization

Even though it is quantionic, Equation (16.19) is still the classical equation of continuity.

We note that the qualifier "quantionic" does not imply "quantum mechanical". It only means "written in terms of quantions". A quantionic expression is still classical if every quantion is multiplicatively paired to its Hermitian conjugate — as Q is with $Q^\dagger$ in the expression (16.3), or $|q)$ is with $(q|$ in the expression (16.19).

Structural quantization yields a quantum mechanical equation by separating the quantionic function from its Hermitian conjugate. The equation of continuity (16.19) lends itself particularly well to such a separation. Indeed, the solution suggests itself if the problem is stated as follows:

Writing

$$
\mathcal{D}\,|q) = |X), \qquad (16.20)
$$

find the most general algebraic solution for $|X)$ which guarantees that $(q|X)$ is purely imaginary.

Setting aside the question of self-consistency, which is the subject matter of Chapter 17, the general solution is given by the next theorem. But a reminder is needed to prevent misunderstanding: The complex conjugate $|q^*)$ of $|q)$ is not to be confused with its Hermitian conjugate $(q|$. Thus:

$$(q| = \begin{pmatrix} q_1^* & q_2^* & q_3^* & q_4^* \end{pmatrix},$$
$$(q^*| = \begin{pmatrix} q_1 & q_2 & q_3 & q_4 \end{pmatrix},$$

and

$$|q^*) = \begin{pmatrix} q_1^* \\ q_2^* \\ q_3^* \\ q_4^* \end{pmatrix}, \quad |q) = \begin{pmatrix} q_1 \\ q_2 \\ q_3 \\ q_4 \end{pmatrix}.$$

Theorem 63 *The most general **quantionic field equation** is of the form*

$$\mathcal{D}|q) = iH|q) + iA|q^*), \tag{16.21}$$

where H is a 4×4 Hermitian matrix and A a complex 4×4 antisymmetric matrix.

Proof. There are two special cases for which Equation (16.18) is identically satisfied:

$$(q|\mathcal{D}|q) \equiv \text{ imaginary} \neq 0, \tag{16.22}$$

and

$$(q|\mathcal{D}|q) \equiv 0. \tag{16.23}$$

The most general linear differential equation which guarantees the condition (16.22) is

$$\mathcal{D}|q) = iH|q), \tag{16.24}$$

where H is an arbitrary 4×4 Hermitian matrix. This is evident.

The most general linear differential equation which guarantees the condition (16.23) is

$$\mathcal{D}|q) = iA|q^*), \tag{16.25}$$

where A is an arbitrary complex 4×4 antisymmetric matrix. While essential in (16.24), the imaginary factor i plays only an aesthetic role in this equation.

To see why the complex conjugate is needed, let us write the expression $(q|\,A\,|q^*)$ in components:

$$(q|\,A\,|q^*) = \sum_{i,j=1}^{4} q_i^* A_{ij} q_j^*.$$

This expression vanishes identically if and only if the matrix A is antisymmetric.

As the sum of the special cases (16.24) and (16.25), Equation (16.21) is the most general expression for $|\mathcal{D}q)$. ∎

Being of first order in the components of the quantionic fields $|q\,(x))$ and $|q^*\,(x))$, Equation (16.21) is no longer classical, but quantum mechanical. We refer to it as 'general' because the matrices H and A are not yet fully specified. They must be Hermitian and antisymmetric respectively, but these are only necessary conditions.

Equation (16.21) suggests the following terminology for the 'separating matrices' H and A, which are analogous in their roles to the 'separating functions' F and N in the derivation of the Schrödinger equation — see relations (15.9) and (15.10).

The matrix H will be referred to as the **matrix potential.** The intuitive reason is that the operator $(\mathcal{D} - iH)$ may be interpreted as a covariant derivative, so that H plays the role of a differential connection, or potential, depending on the point of view.

The matrix A will be written in the form $A = mM$, where m is the mass of the field. The dimensionless matrix M will be referred to as the **mass matrix.** The intuitive reason is that the field $|q^*)$ plays the role of the 'source' in Equation (16.21), while A is the coupling coefficient — which is how the mass enters in all spinorial field equations.

Chapter 17

The Mass Matrix

It is heuristically evident that the coupling matrix A in the quantionic field equation (16.21) characterizes the mass of the quantionic field. The present chapter is dedicated to the analysis and simplification of this matrix.

All we know at this point concerning the matrix A is that it is complex and antisymmetric. It is thus uniquely defined by 12 free real parameters. We show in this chapter that only one parameter, interpreted as the field mass m, is physically meaningful. The proof consists of three steps illustrated in the following diagram. It begin with an arbitrary complex antisymmetric matrix and terminate with Dirac's matrix γ^1.

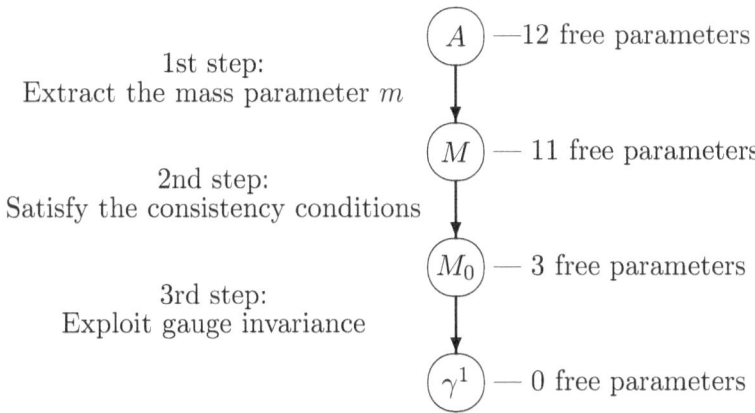

1st step:
Extract the mass parameter m

A —12 free parameters

2nd step:
Satisfy the consistency conditions

M — 11 free parameters

3rd step:
Exploit gauge invariance

M_0 — 3 free parameters

γ^1 — 0 free parameters

One section of the present chapter is dedicated to each step.

In the first step, Section 17.1, we extract from A a uniquely defined parameter m,

$$A = mM. \tag{17.1}$$

As in the Schrödinger and Dirac equations, m is interpreted as the mass of the quantionic field. The dimensionless matrix M will be referred to as the **mass matrix.**

In the second step, Section 17.2, we show that the mass matrix M is defined by only three parameters, not eleven. The reduction is forced by a consistency condition.

In the third step, Section 17.3, we show that these three parameters are not physically observable because they can be eliminated by a gauge transformation. This leaves only one physically meaningful parameter — the mass m. It follows that the simplest form of the matrix A is

$$A = m\gamma^1, \tag{17.2}$$

where

$$\gamma^1 = \begin{pmatrix} 0 & \sigma_1 \\ -\sigma_1 & 0 \end{pmatrix}$$

is one of Dirac's gamma matrices. As require of A, the matrix γ^1 is antisymmetric.

17.1   The mass parameter

The following two observations lead to the first simplification of the matrix A :

(1) The matrix elements of the derivation operator $\mathcal{D}$ are homogeneous in the differential operators ∂_μ. The dimension of the gradient ∇ being a reciprocal length, the standard conventions $c = \hbar = 1$ imply that $\{\partial_\mu\}$ has the dimension of reciprocal length, or mass.

(2) Since the fields $|q\rangle$ and $|q^*\rangle$ have the same dimension (regardless of what it is), the field equation (16.21) implies that the matrix elements of A must have the dimension of mass as well.[1]

---

[1] The same is true for the matrix elements of H, but this is irrelevant in the present chapter because we assumed $H = 0$.

It is thus natural to factor out of A a mass parameter m, as in relation (17.1). This leaves eleven free dimensionless parameters in the matrix M. But a normalization condition is needed for such a factorization to be uniquely defined. The following condition suggests itself as most natural:

$$\|\det M\|^2 \overset{def}{=} (\det M)^* (\det M) = 1. \tag{17.3}$$

Equivalently,

$$\det M = e^{i4\chi}. \tag{17.4}$$

Since, in general, $\det M$ is a complex number, one cannot impose the simpler normalization condition $\det M = 1$ because it would incorrectly eliminate from M two parameters instead of one. The factor 4 in the exponent is aesthetic: It makes it possible to factor out of the matrix M the phase factor $e^{i\chi}$, which is more elegant than $e^{i\chi/4}$.

The field equation (16.21) now reads

$$\mathcal{D}\,|q) = imM\,|q^*)\,. \tag{17.5}$$

At this point, the matrix M is defined only as a complex antisymmetric matrix that satisfies the real condition (17.3). It thus consists of eleven free real parameters that could be spacetime functions. In the present chapter, we consider only the underlying algebraic aspect of consistency, which means that the matrix elements may be constants. Hence, their spacetime derivatives vanish:

$$(\mathcal{D}M) = 0. \tag{17.6}$$

The general case, which involves local gauge transformations, will be investigated separately.

17.2   The consistency conditions

The fields $|q)$ and $|q^*)$ are related by two independent conditions: An algebraic condition (complex conjugation) and a differential condition (the differential equation (17.5)). In general, these conditions are not mutually consistent. The consistency requirement is the subject of the following theorem.

Theorem 64 *The most general mass matrix M in the field equation (17.5) is of the form*

$$M = \begin{pmatrix} 0 & N \\ -N & 0 \end{pmatrix}, \tag{17.7}$$

where N is a 2×2 matrix of the form

$$N = e^{i\chi} \begin{pmatrix} z & r \\ r & -z^* \end{pmatrix}, \tag{17.8}$$

whose elements $r \in \mathbb{R}$ and $z \in \mathbb{C}$ are normalized by the condition

$$z^* z + r^2 = 1. \tag{17.9}$$

Proof. The proof consists of two parts:

(1) Derive from Equation (17.5) two equivalent equations, and, by mixing them, eliminate either the field $|q\rangle$ or the field $|q^*\rangle$, whichever falls out with least effort. This yields an identity in the field that has not been eliminated.

(2) Such an identity in the remaining field is an equation relating the operators $\mathcal{D}$ and M which act on that field. Since the matrix operator $\mathcal{D}$ is known, one can compute the most general matrix M from the equation that relates it to the matrix $\mathcal{D}$.

First part of the proof:

Complex conjugation yields the first equation equivalent to Equation (17.5):

$$\mathcal{D}^* |q^*\rangle = -imM^* |q\rangle. \tag{17.10}$$

One obtains the second equation by applying the operator $\mathcal{D}^{\#}$ to both sides of Equation (17.5). By theorem 62, the differential operator $\mathcal{D}^{\#}\mathcal{D}$ is the D'Alambertian, so that

$$im\mathcal{D}^{\#} M |q^*\rangle = \square |q\rangle. \tag{17.11}$$

Since the D'Alambertian

$$\square = D\Delta - \delta\delta^*$$

is a scalar operator, it commutes with all constant matrices and with all differential operators. Thus, by first applying the D'Alambertian to both sides of Equation (17.10), and then multiplying Equation (17.11) by $-imM^*$, one obtains the following two equations, whose right-hand sides are identical:

$$\mathcal{D}^*\Box\,|q^*) \;=\; -imM^*\Box\,|q)\,,$$
$$m^2 M^* \mathcal{D}^\# M\,|q^*) \;=\; -imM^*\Box\,|q)\,.$$

It follows that the left-hand sides of these equations are identical for all solutions $|q)$ of the field Equation (17.5):

$$\mathcal{D}^*\Box\,|q^*) \equiv m^2 M^* \mathcal{D}^\# M\,|q^*)\,. \tag{17.12}$$

By applying the operator $\mathcal{D}^{\#*}$ to both sides of this equation and by taking into account the relations

$$\mathcal{D}^{\#*}\mathcal{D}^* = \left(\mathcal{D}^\#\mathcal{D}\right)^* = \Box^* = \Box,$$

one eliminate the operator $\mathcal{D}^*$ from the left-hand side:

$$\Box^2\,|q^*) \equiv m^2 \mathcal{D}^{\#*} M^* \mathcal{D}^\# M\,|q^*)\,.$$

By introducing a new temporary matrix operator,

$$K \stackrel{def}{=} \mathcal{D}^\# M\,, \tag{17.13}$$

we formally simplify the previous equation to

$$\Box^2\,|q^*) = m^2 K^* K\,|q^*)\,. \tag{17.14}$$

This identity in the field $|q^*)$ is the defining equation for the matrix K, and hence for the mass matrix M.

Second part of the proof:

To solve Equation (17.14), let us write the unknown antisymmetric matrix M in block form,

$$M = \begin{pmatrix} R & N \\ -N^\top & S \end{pmatrix}, \tag{17.15}$$

where R and S are complex antisymmetric 2×2 matrices and N is an arbitrary complex 2×2 matrix. Similarly, the derivation operators in block matrix form read:

$$\left.\begin{aligned} \mathcal{D} &= \begin{pmatrix} D & \delta \\ \delta^* & \Delta \end{pmatrix}, \\ \mathcal{D}^\# &= \begin{pmatrix} \Delta & -\delta \\ -\delta^* & D \end{pmatrix}. \end{aligned}\right\} \tag{17.16}$$

The 2×2 unit matrix I has been dropped because it is not needed formally. Conceptually, however, the matrix elements should be thought of as DI, ΔI, and δI.

Substitution of the matrices (17.15) and (17.16) into the definition (17.13) of the matrix K yields

$$\begin{aligned} K &= \begin{pmatrix} \Delta & -\delta \\ -\delta^* & D \end{pmatrix} \begin{pmatrix} R & N \\ -N^\mathsf{T} & S \end{pmatrix} \\ &= \begin{pmatrix} R\Delta + N^\mathsf{T}\delta & N\Delta - S\delta \\ -R\delta^* - N^\mathsf{T}D & SD - N\delta^* \end{pmatrix}, \end{aligned}$$

and

$$K^* = \begin{pmatrix} R^*\Delta + N^{*\mathsf{T}}\delta^* & N^*\Delta - S^*\delta^* \\ -R^*\delta - N^{*\mathsf{T}}D & S^*D - N^*\delta \end{pmatrix}.$$

Let us first compute the off-diagonal term $(K^*K)_{12}$, which, by Equation (17.14), must vanish:

$$\begin{aligned} &(K^*K)_{12} \\ &= \left(R^*\Delta + N^{*\mathsf{T}}\delta^*\right)(N\Delta - S\delta) + (N^*\Delta - S^*\delta^*)(SD - N\delta^*) \\ &= \left(R^*N\Delta\Delta + N^{*\mathsf{T}}N\delta^*\Delta\right) - \left(R^*S\Delta\delta + N^{*\mathsf{T}}S\delta^*\delta\right) \\ &\quad + (N^*S\Delta D - S^*S\delta^*D) - (N^*N\Delta\delta^* - S^*N\delta^*\delta^*) \\ &= R^*N\Delta^2 - R^*S\Delta\delta + N^*S\Delta D - S^*S\delta^*D \\ &\quad + \left(N^{*\mathsf{T}}N - N^*N\right)\Delta\delta^* + \left(N^{*\mathsf{T}}S - S^*N\right)\delta^*\delta^*. \end{aligned}$$

This expression vanishes if and only if all coefficients of the differential operators vanish, which implies:

$$R^*N = R^*S = N^*S = S^*S = 0,$$
$$-N^{*\top}S + S^*N = 0,$$
$$N^{*\top}N - N^*N = 0.$$

The identity $(K^*K)_{21} = 0$ yields the same conditions, but with R substituted for S. It follows that N is symmetric,

$$N^\top = N$$

and that

$$R = S = 0.$$

These results simplify the mass matrix (17.15) to the form

$$M = \begin{pmatrix} 0 & N \\ -N & 0 \end{pmatrix}. \tag{17.17}$$

The matrices K and K^* now read:

$$K = \begin{pmatrix} N\delta & N\Delta \\ -ND & -N\delta^* \end{pmatrix},$$
$$K^* = \begin{pmatrix} N^*\delta^* & N^*\Delta \\ -N^*D & -N^*\delta \end{pmatrix}.$$

Hence:

$$K^*K = \begin{pmatrix} N^*\delta^* & N^*\Delta \\ -N^*D & -N^*\delta \end{pmatrix} \begin{pmatrix} N\delta & N\Delta \\ -ND & -N\delta^* \end{pmatrix}$$
$$= [N^*N(\delta\delta^* - \Delta D)]I = -N^*N\Box I.$$

Substitution of these expressions into relation (17.14) yields

$$\Box^2 |q^*) = -m^2 N^*N\Box |q^*),$$

which simplifies to

$$\Box |q^*) = -m^2 N^*N |q^*). \tag{17.18}$$

Since the D'Alambertian is a scalar operator, it follows that N^*N must be proportional to the unit matrix:

$$N^*N = \lambda I. \tag{17.19}$$

Let us write the symmetric matrix N in the form

$$N = e^{i\chi} \begin{pmatrix} z & r \\ r & v \end{pmatrix},$$

where r is a real number while z and v are complex numbers. The condition (17.19) yields

$$
\begin{aligned}
N^*N &= \begin{pmatrix} z^* & r \\ r & v^* \end{pmatrix} \begin{pmatrix} z & r \\ r & v \end{pmatrix} \\
&= \begin{pmatrix} r^2 + zz^* & rz^* + rv \\ rv^* + rz & r^2 + vv^* \end{pmatrix} = \lambda \begin{pmatrix} 1 & 0 \\ 0 & 1 \end{pmatrix},
\end{aligned}
$$

which implies

$$
\begin{aligned}
v &= -z^*, \\
\lambda &= r^2 + zz^* > 0.
\end{aligned}
$$

This reduces the matrix N to

$$N = e^{i\chi} \begin{pmatrix} z & r \\ r & -z^* \end{pmatrix}$$

The matrix M is is now given by relation (17.17):

$$
M = e^{i\chi} \begin{pmatrix} 0 & 0 & z & r \\ 0 & 0 & r & -z^* \\ -z & -r & 0 & 0 \\ -r & z^* & 0 & 0 \end{pmatrix}.
$$

Only λ remains to be determined. The condition (17.4) implies

$$
\det \begin{pmatrix} 0 & 0 & z & r \\ 0 & 0 & r & -z^* \\ -z & -r & 0 & 0 \\ -r & z^* & 0 & 0 \end{pmatrix} = \left(r^2 + z^*z\right)^2 = 1.
$$

Hence, $\lambda = 1$, and
$$r^2 + zz^* = 1,$$
which proves the theorem. ■

This completes the second step in the program of simplification outlined on page 17.

We observe, incidentally, that this proof contains an additional conclusion:

Corollary 65 *Each of the four components of the field $|q)$ satisfies the same Klein-Gordon equation*

$$\Box |q) = -m^2 |q) . \qquad (17.20)$$

Proof. Substitution of $N^*N = I$ into the complex conjugate of relation (17.18) yields Equation (17.20). ■

17.3   The equivalence of all mass matrices

As derived in Theorem 64, the mass matrix M is not of a familiar type, but it can be re-written in a form which relates it to gauge transformations.

Theorem 66 *The most general mass matrix M may be written in the form*

$$M = \gamma^1 F, \qquad (17.21)$$

where $F \in \mathcal{L}$ is a quantionic phase factor,

$$F^\dagger F = I, \qquad (17.22)$$

subject to the condition

$$Tr\,(F\Lambda_3) = 0. \qquad (17.23)$$

Proof. Let us consider a very special L-type quantion of the form

$$F = e^{i\chi} F_0 = e^{i\chi} \begin{pmatrix} r & -z^* & 0 & 0 \\ z & r & 0 & 0 \\ 0 & 0 & r & -z^* \\ 0 & 0 & z & r \end{pmatrix} \in \mathcal{L}. \qquad (17.24)$$

Then:

$$
\gamma^1 F \;=\; e^{i\chi}
\begin{pmatrix}
0 & 0 & 0 & 1 \\
0 & 0 & 1 & 0 \\
0 & -1 & 0 & 0 \\
-1 & 0 & 0 & 0
\end{pmatrix}
\begin{pmatrix}
r & -z^* & 0 & 0 \\
z & r & 0 & 0 \\
0 & 0 & r & -z^* \\
0 & 0 & z & r
\end{pmatrix}
$$

$$
\;=\; e^{i\chi}
\begin{pmatrix}
0 & 0 & z & r \\
0 & 0 & r & -z^* \\
-z & -r & 0 & 0 \\
-r & z^* & 0 & 0
\end{pmatrix}
= M,
$$

which verifies relation (17.21).

It remains to be shown that the matrices F of type (17.24) are characterized by the conditions (17.22) and (17.23).

To this end, consider an arbitrary phase factor F, as given by relation (9.17). Since the scalar factor $e^{i\chi}$ commutes with all matrices, we may ignore it. We thus have

$$
F_0 \Lambda_3 = \begin{pmatrix} U & 0 \\ 0 & U \end{pmatrix} \begin{pmatrix} \sigma_3 & 0 \\ 0 & \sigma_3 \end{pmatrix} = \begin{pmatrix} U\sigma_3 & 0 \\ 0 & U\sigma_3 \end{pmatrix}.
$$

Thus, the trace of $F_0\Lambda_3$ vanishes if and only if the trace of $U\sigma_3$ vanishes. Explicitly:

$$
\begin{aligned}
U\sigma_3 \;&=\; \begin{pmatrix} \cos\gamma\, e^{i\alpha} & \sin\gamma\, e^{i\beta} \\ -\sin\gamma\, e^{-i\beta} & \cos\gamma\, e^{-i\alpha} \end{pmatrix} \begin{pmatrix} 1 & 0 \\ 0 & -1 \end{pmatrix} \\
&=\; \begin{pmatrix} e^{i\alpha}\cos\gamma & -e^{i\beta}\sin\gamma \\ -e^{-i\beta}\sin\gamma & -e^{-i\alpha}\cos\gamma \end{pmatrix}
\end{aligned}
$$

Hence,

$$
Tr\,(U\sigma_3) = 2i\sin\alpha\cos\gamma = 0,
$$

which implies $\alpha = 0$ or $\gamma = \pm\frac{\pi}{2}$.

For $\alpha = 0$, the matrix F is indeed of the form (17.24), the identifications being

$$
\left.\begin{aligned}
r &= \cos\gamma, \\
z &= -\sin\gamma\, e^{-i\beta}.
\end{aligned}\right\} \tag{17.25}
$$

For $\gamma = \pm\frac{\pi}{2}$, the solution is merely a special case of the general solution (17.25). This completes the proof of the theorem. ∎

It is interesting to note that the two special numerical matrices γ^1 and Λ_3 play distinguished roles in the structure of the mass matrix. We shall see later that they are also essential in the classification of the matrix potentials H. The following lemma shows that these matrices are mutually related by way of the concept of a quantionic phase factor.

Lemma 67 *If F is a quantionic phase factor, that is, an L-type quantion such such that $F^\dagger F = I$, the conditions*

$$Tr\,(F\Lambda_3) = 0$$

and

$$F\gamma^1 F^* = \gamma^1 \qquad (17.26)$$

are equivalent.

Proof. Dropping the factor $e^{i\chi}$, which cancels in both relations, it suffices to consider the pure quantionic phase factor

$$F_0 = \begin{pmatrix} U & 0 \\ 0 & U \end{pmatrix}.$$

In the parametrization

$$U = \begin{pmatrix} \cos\gamma e^{i\alpha} & \sin\gamma e^{i\beta} \\ -\sin\gamma e^{-i\beta} & \cos\gamma e^{-i\alpha} \end{pmatrix},$$

we have

$$\begin{aligned}
F\gamma^1 F^* &= \begin{pmatrix} U & 0 \\ 0 & U \end{pmatrix} \begin{pmatrix} 0 & \sigma_1 \\ -\sigma_1 & 0 \end{pmatrix} \begin{pmatrix} U^* & 0 \\ 0 & U^* \end{pmatrix} \\
&= \begin{pmatrix} 0 & U\sigma_1 \\ -U\sigma_1 & 0 \end{pmatrix} \begin{pmatrix} U^* & 0 \\ 0 & U^* \end{pmatrix} \\
&= \begin{pmatrix} 0 & U\sigma_1 U^* \\ -U\sigma_1 U^* & 0 \end{pmatrix}.
\end{aligned}$$

Hence, the relation $F\gamma^1 F^* = \gamma^1$ is equivalent to

$$U\sigma_1 U^* = \sigma_1. \qquad (17.27)$$

Explicitly:

$$U\sigma_1 = \begin{pmatrix} \cos\gamma e^{i\alpha} & \sin\gamma e^{i\beta} \\ -\sin\gamma e^{-i\beta} & \cos\gamma e^{-i\alpha} \end{pmatrix}\begin{pmatrix} 0 & 1 \\ 1 & 0 \end{pmatrix}$$

$$= \begin{pmatrix} e^{i\beta}\sin\gamma & e^{i\alpha}\cos\gamma \\ e^{-i\alpha}\cos\gamma & -e^{-i\beta}\sin\gamma \end{pmatrix},$$

$$U\sigma_1 U^* = \begin{pmatrix} e^{i\beta}\sin\gamma & e^{i\alpha}\cos\gamma \\ e^{-i\alpha}\cos\gamma & -e^{-i\beta}\sin\gamma \end{pmatrix}\begin{pmatrix} \cos\gamma e^{-i\alpha} & \sin\gamma e^{-i\beta} \\ -\sin\gamma e^{i\beta} & \cos\gamma e^{i\alpha} \end{pmatrix}$$

$$= \begin{pmatrix} i\sin\alpha\sin2\gamma e^{i\beta} & e^{i2\alpha}\cos^2\gamma + \sin^2\gamma \\ e^{-i2\alpha}\cos^2\gamma + \sin^2\gamma & -i\sin\alpha\sin2\gamma e^{-i\beta} \end{pmatrix}.$$

Relation (17.27) is equivalent to $\alpha = k\pi$. Taking k odd is equivalent to renaming γ, namely $\gamma \to \pi - \gamma$. The general solution is thus $\alpha = 0$, which proves the assertion. ∎

In the notation introduced in Section 9.3, the matrix F is of the type $F = e^{i\chi}F_0$, where

$$F_0 = \begin{pmatrix} U(\gamma, 0, \beta) & 0 \\ 0 & U(\gamma, 0, \beta) \end{pmatrix}. \tag{17.28}$$

Hence, by Theorem 39, the matrices F_0 do not form a group.

Substitution of the expression (17.21) for M into the field Equation (17.5) yields

$$\mathcal{D}|q\rangle = im\gamma^1 F|q^*\rangle. \tag{17.29}$$

Let us now take a particular 'square root' of F by defining a matrix G such that

$$G^*G^* = F. \tag{17.30}$$

For F given by the expression (17.24), it is easy to verify that the solution for G is

$$G = e^{-i\chi/2}\begin{pmatrix} s & -u & 0 & 0 \\ u^* & s & 0 & 0 \\ 0 & 0 & s & -u \\ 0 & 0 & u^* & s \end{pmatrix}, \tag{17.31}$$

where the new parameters

$$\left.\begin{array}{l} s = \sqrt{\frac{r+1}{2}}, \\ u = \frac{z^*}{\sqrt{2(r+1)}}, \end{array}\right\} \tag{17.32}$$

satisfy the normalization condition

$$s^2 + u^*u = 1.$$

Note: This proof is only meant to show that the matrix G is uniquely defined by the matrix F (up to an irrelevant sign) and that it satisfies the same conditions as F itself. For computations, however, the parametric form of the matrix G is more convenient than the algebraic expression (17.31).

Substitution of relation (17.30) into the field Equation (17.29) yields the equivalent equation

$$\mathcal{D}\,|q) = im\gamma^1 G^* G^*\,|q^*)\,.$$

Multiplying both sides of this equation on the left by G yields

$$G\mathcal{D}\,|q) = imG\gamma^1 G^* G^*\,|q^*)\,. \tag{17.33}$$

This equation admits three simplifications:

(1) Since G and $\mathcal{D}$ are L-type and R-type quantions respectively, they commute algebraically, and since the parameters β, γ, and χ are assumed to be constants, they also commute differentially. Thus, $G\mathcal{D} = \mathcal{D}G$.

(2) Relation (17.26) implies $G\gamma^1 G^* = \gamma^1$.

(3) If we denote by $|\tilde{q})$ the field related to $|q)$ by the gauge transformation G, that is:

$$|\tilde{q}) \stackrel{def}{=} G\,|q)\,, \tag{17.34}$$

then

$$G^*\,|q^*) = |\tilde{q}^*)\,.$$

Hence, the field Equation (17.33) simplifies to

$$\mathcal{D}\,|\tilde{q}) = im\gamma^1\,|\tilde{q}^*)\,, \tag{17.35}$$

which is formally the same as Equation (17.5).

Conclusion 68 *The three parameters β, γ, and χ in the most general mass matrix M can be eliminated by a gauge transformation of the quantionic field. Consequently, they have no observable physical meaning. After their elimination, all formally different mass matrices reduce to the same form,*

$$M = \gamma^1. \tag{17.36}$$

*We shall refer to this maximally reduced form of the mass matrix as the **Dirac gauge**.*

This terminology is not to suggest that these ideas go back to Dirac. Its justification is that taking $M = \gamma^1$ in the quantionic field equation yields the Dirac equation. This is shown in the next chapter.

The special mass matrix γ^1 is not only convenient for its simplicity, it is also structurally distinguished, as illustrated in the following diagram:

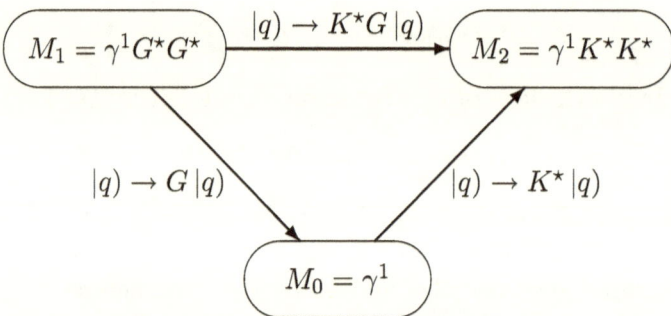

In this commutative diagram, the mass matrices M_1 and M_2 are arbitrary. Both are related to the matrix γ^1 by operators that satisfy the condition (17.26), but they are also related among themselves by an operator that does not satisfy the same condition: Formally, the gauge transformation $|q) \mapsto G\,|q)$ followed by the gauge transformation $|q) \mapsto K^*\,|q)$ maps the matrix M_1 to γ^1, and then γ^1 to M_2. Hence, the composite transformation $|q) \mapsto K^*G\,|q)$ maps M_1 to M_2. But while the matrices G and K^* satisfy the condition (17.26), their product K^*G does not because, in general,

$$(K^*G)^* \, \gamma^1 \, (K^*G) \neq \gamma^1.$$

This concludes the program outlined on page 207.

Chapter 18

The Quantionic Dirac Equation

An essential result of the last two chapters is that the mass matrix M contains no free parameter. Let us emphasize this point:

The most general quantionic field equation is (16.21), but all complex antisymmetric matrices A which are also consistent with the complex conjugate of this equation are related to the specific matrix $m\gamma^1$ by quantionic gauge transformations $E \in U_q(1)$.

This result implies that the quantionic field equation written in the Dirac gauge,

$$[\mathcal{D} - iH]\,|q) = im\gamma^1\,|q^*)\,, \tag{18.1}$$

may be taken, without loss of generality, as the starting point of all subsequent investigations.

In the first section of this chapter, we shall prove that the four component submatrices H_α in the decomposition (4.27) of the general matrix potential,

$$H = R^\mu\Lambda_\mu = H_0 + H_1 + H_2 + H_3,$$

are differential gauge connections for the four Abelian subgroups of the gauge group $U_q(1)$. In the exponential expression (9.18) for the general element $E(x) \in U_q(1)$, we have:

$$
\begin{aligned}
H_0 &= R\Lambda_0 \longleftrightarrow E_0 = \exp(i\phi\Lambda_0), \\
H_1 &= R\Lambda_1 \longleftrightarrow E_1 = \exp(i\phi\Lambda_1), \\
H_2 &= R\Lambda_2 \longleftrightarrow E_2 = \exp(i\phi\Lambda_2), \\
H_3 &= R\Lambda_3 \longleftrightarrow E_3 = \exp(i\phi\Lambda_3),
\end{aligned}
$$

where ϕ stands for a scalar field, $\phi = \phi(x)$.

In the historical development of physics, covariant derivatives came to light in two steps: Ordinary derivative were initially introduced in a fixed basis, or gauge; much later, they were made covariant with respect to newly identified groups of relevant transformations. This first took place in general relativity, and then in gauge theory. In contrast, we shall see that *structural quantization gives rise directly to the derivation operator D in a form covariant with respect to the group $U_q(1)$.*

One of the four Abelian subgroups of $U_q(1)$ — the group of transformations $E_3 = \exp(i\phi\Lambda_3)$ — is distinguished by leaving the mass matrix invariant. In the second subsection, we take advantage of this property to extract from the general equation (18.1) a special equation which is gauge covariant. We refer to it as the "quantionic Dirac equation". This terminology is justified in Chapter 19, where a non-trivial equivalence is established between quantions and Dirac four-spinors.

18.1   Gauge connections and potentials

A general gauge transformation $E \in U_q(1)$ transforms the quantionic fields according to the relations

$$
\left.
\begin{aligned}
|q) &\longmapsto |q') = E\,|q), \\
|q^*) &\longmapsto |q'^*) = E^*\,|q^*).
\end{aligned}
\right\}
\tag{18.2}
$$

Substitution of these expressions into Equation (18.1) yields

$$
\mathcal{D}E\,|q) - iHE\,|q) = im\gamma^1 E^*\,|q^*).
\tag{18.3}
$$

For a local gauge transformation, the unitary matrix E is a space-time function, so that, in general, $(\mathcal{D}E) \neq 0$. This causes no conceptual difficulty because the Leibniz identity

$$\mathcal{D}E\,|q) = E\mathcal{D}\,|q) + (\mathcal{D}E)\,|q) \qquad (18.4)$$

is guaranteed: E is an L-type quantion, while $\mathcal{D}$ is an R-type matrix — which implies algebraic commutativity.

The substitution of this expression for $\mathcal{D}E\,|q)$ into Equation (18.3) yields

$$E\mathcal{D}\,|q) - i\,[HE + i\,(\mathcal{D}E)]\,|q) = im\gamma^1 E^*\,|q^*)\,. \qquad (18.5)$$

Multiplying this equation from the left by $E^\dagger$ brings it formally closer to the reference equation (18.1):

$$\mathcal{D}\,|q) - i\left[E^\dagger HE + iE^\dagger\,(\mathcal{D}E)\right]|q) = imE^\dagger\gamma^1 E^*\,|q^*)\,. \qquad (18.6)$$

We are to investigate how the new mass matrix

$$M' = E^\dagger\gamma^1 E^* \qquad (18.7)$$

and the new potential

$$H' = E^\dagger HE + iE^\dagger\,(\mathcal{D}E) \qquad (18.8)$$

are related to the subgroups of the gauge group $U_q\,(1)$.

It is convenient for computations to write the transformed matrix H' as the sum of an 'algebraic part' H'_a and of a 'differential part' H'_d :

$$H' = H'_a + H'_d, \qquad (18.9)$$

where

$$H'_a \overset{def}{=} E^\dagger HE, \qquad (18.10)$$
$$H'_d \overset{def}{=} iE^\dagger\,(\mathcal{D}E)\,.$$

To compute M' and H' in a uniform formalism, we shall expand the Hermitian matrix H and the unitary matrix E in the same basis of lambda matrices.

The lambda-decomposition of the potential matrix is given in Section 4.4, Relation (4.27), that is,

$$H = R^\mu \Lambda_\mu = R^0 \Lambda_0 + R^1 \Lambda_1 + R^2 \Lambda_2 + R^3 \Lambda_3, \qquad (18.11)$$

where R^0 to R^3 are arbitrary R-type quantions.

The lambda-decomposition of the unitary matrix is given in Section 9.3, Relation (9.18), that is,

$$E = e^{i\chi} \exp\left(i\phi\vec{m} \cdot \vec{\Lambda}\right) \in U_q\left(1\right). \qquad (18.12)$$

We shall not work with this general expression but with its four Abelian one-parametric subgroups

$$E_\omega = \cos\phi_\omega\, I + i\sin\phi_\omega\, \Lambda_\omega \in U_\omega\left(1\right). \qquad (18.13)$$

The symbols E_ω and $U_\omega\left(1\right)$ are introduced for uniformity of formalism. By comparison with the previously used symbols, we get $\phi_0 = \chi$ and $U_0\left(1\right) = U\left(1\right)$, while for $i = 1, 2, 3$, we have $\phi_i \overset{def}{=} \phi m_i$. The 'unusual' index ω is chosen as a reminder that it is a label, not a tensor index — even though it runs from 0 to 3.

Taking the four subgroups $U_\omega\left(1\right)$ in turn, we shall compute the corresponding matrices H' and M' according to the general expressions (18.7), (18.10) and (18.11).

Since some identities involving the matrices γ^1 and Λ_μ will be needed repeatedly, it is most convenient to prove them first in several lemmas:

Lemma 69 *The commutation relations of γ^1 and Λ_μ :*

$$\begin{aligned}
\gamma^1 \Lambda_0 &= \Lambda_0 \gamma^1, \\
\gamma^1 \Lambda_1 &= \Lambda_1 \gamma^1, \\
\gamma^1 \Lambda_2 &= -\Lambda_2 \gamma^1, \\
\gamma^1 \Lambda_3 &= -\Lambda_3 \gamma^1.
\end{aligned}$$

Proof. For $\gamma^1 \Lambda_3$, taken as an example, we have

$$\gamma^1 \Lambda_3 = \begin{pmatrix} 0 & \sigma_1 \\ -\sigma_1 & 0 \end{pmatrix} \begin{pmatrix} \sigma_3 & 0 \\ 0 & \sigma_3 \end{pmatrix} = \begin{pmatrix} 0 & -i\sigma_2 \\ i\sigma_2 & 0 \end{pmatrix}$$

$$\Lambda_3 \gamma^1 = \begin{pmatrix} \sigma_3 & 0 \\ 0 & \sigma_3 \end{pmatrix} \begin{pmatrix} 0 & \sigma_1 \\ -\sigma_1 & 0 \end{pmatrix} = \begin{pmatrix} 0 & i\sigma_2 \\ -i\sigma_2 & 0 \end{pmatrix}$$

and similarly for the other two cases. ∎

These relations imply the following commutation relations for the matrices γ^1 and E_μ :

Lemma 70 *The commutation relations of γ^1 and E_μ :*

$$\gamma^1 E_0^* = E_0^* \gamma^1,$$
$$\gamma^1 E_1^* = E_1^* \gamma^1,$$
$$\gamma^1 E_2^* = E_2^* \gamma^1,$$
$$\gamma^1 E_3^* = E_3 \gamma^1.$$

Proof. For E_0^*, the relation is evident.

$$
\begin{aligned}
\gamma^1 E_1^* &= \gamma^1 \left(\cos\phi\, I - i\sin\phi\, \Lambda_1\right) = \left(\cos\phi\, I\gamma^1 - i\sin\phi\, \gamma^1\Lambda_1\right) \\
&= \left(\cos\phi\, I\gamma^1 - i\sin\phi\, \Lambda_3\gamma^1\right) = E_1^*\gamma^1,
\end{aligned}
$$

$$
\begin{aligned}
\gamma^1 E_2^* &= \gamma^1 \left(\cos\phi\, I + i\sin\phi\, \Lambda_2\right) = \left(\cos\phi\, I\gamma^1 + i\sin\phi\, \gamma^1\Lambda_2\right) \\
&= \left(\cos\phi\, I\gamma^1 - i\sin\phi\, \Lambda_2\gamma^1\right) = E_2^*\gamma^1,
\end{aligned}
$$

$$
\begin{aligned}
\gamma^1 E_3^* &= \gamma^1 \left(\cos\phi\, I - i\sin\phi\, \Lambda_3\right) = \left(\cos\phi\, I\gamma^1 - i\sin\phi\, \gamma^1\Lambda_3\right) \\
&= \left(\cos\phi\, I\gamma^1 + i\sin\phi\, \Lambda_3\gamma^1\right) = E_3\gamma^1.
\end{aligned}
$$

We see that the last case differs from the other three. This seems to be related to the unlimited reach of electromagnetic interactions. ∎

The proofs of the following two lemmas are similarly straightforward.

Lemma 71 *The commutators of lambda matrices:*

$[\Lambda_i, \Lambda_j]$	Λ_1	Λ_2	Λ_3
Λ_1	0	$i2\Lambda_3$	$-i2\Lambda_2$
Λ_2	$-i2\Lambda_3$	0	$i2\Lambda_1$
Λ_3	$i2\Lambda_2$	$-i2\Lambda_1$	0

Lemma 72 *The triple products of lambda matrices:*

$\Lambda_1\Lambda_0\Lambda_1 = \Lambda_0$	$\Lambda_2\Lambda_0\Lambda_2 = \Lambda_0$	$\Lambda_3\Lambda_0\Lambda_3 = \Lambda_0$
$\Lambda_1\Lambda_1\Lambda_1 = \Lambda_1$	$\Lambda_2\Lambda_1\Lambda_2 = -\Lambda_1$	$\Lambda_3\Lambda_1\Lambda_3 = -\Lambda_1$
$\Lambda_1\Lambda_2\Lambda_1 = -\Lambda_2$	$\Lambda_2\Lambda_2\Lambda_2 = \Lambda_2$	$\Lambda_3\Lambda_2\Lambda_3 = -\Lambda_2$
$\Lambda_1\Lambda_3\Lambda_1 = -\Lambda_3$	$\Lambda_2\Lambda_3\Lambda_2 = -\Lambda_3$	$\Lambda_3\Lambda_3\Lambda_3 = \Lambda_3$

We now have all the identities necessary to investigate the actions of the subgroups $U_\omega(1)$. In the following computation, the free labels ω may be dropped for convenience from ϕ^ω and R^ω because the groups $U_1(1)$ to $U_3(1)$ are studied individually.

The group $U_0(1)$

This is the standard gauge group $U(1)$ of complex phase transformations. For

$$E = e^{i\phi} \in U(1),$$

the transformed matrices are

$$M' = e^{-i2\phi}\gamma^1. \tag{18.14}$$

and

$$\begin{aligned} H'_a &= H, \\ H'_d &= -(\mathcal{D}\phi). \end{aligned}$$

Hence

$$H' = H - (\mathcal{D}\phi), \tag{18.15}$$

The transformations of the coefficients R^ω are most conveniently represented in vector form:

$$\begin{pmatrix} R^0 \\ R^1 \\ R^2 \\ R^3 \end{pmatrix} \mapsto \begin{pmatrix} R'^0 \\ R'^1 \\ R'^2 \\ R'^3 \end{pmatrix} = \begin{pmatrix} R^0 - \mathcal{D}\phi \\ R^1 \\ R^2 \\ R^3 \end{pmatrix}. \tag{18.16}$$

The group $U_1(1)$:

The general unitary matrix is

$$E = \cos\phi \, I + i\sin\phi \, \Lambda_1 \in U_1(1).$$

For M', Lemma 70 implies

$$M' = E^\dagger \gamma^1 E^* = E^\dagger E^* \gamma^1 = (\cos\phi \, I - i\sin\phi \, \Lambda_1)^2 \, \gamma^1,$$

which yields

$$M' = e^{-i2\phi \, \Lambda_1} \gamma^1. \tag{18.17}$$

For H', the two components transform according to

$$
\begin{aligned}
H'_a &= (\cos\phi \, I - i\sin\phi \, \Lambda_1) \, H \, (\cos\phi \, I + i\sin\phi \, \Lambda_1), \\
H'_d &= i\,(\cos\phi \, I - i\sin\phi \, \Lambda_1) \, \mathcal{D} \, (\cos\phi \, I + i\sin\phi \, \Lambda_1).
\end{aligned}
$$

Let us compute the algebraic part H'_a (exceptionally, in the following expressions, ω is a summation index, not a free label):

$$
\begin{aligned}
H'_a &= (\cos\phi \, I - i\sin\phi \, \Lambda_1) \, R^\omega \Lambda_\omega \, (\cos\phi \, I + i\sin\phi \, \Lambda_1) \\
&\equiv \cos^2\phi \, H + i\cos\phi\sin\phi \, R^\omega \, [\Lambda_\omega, \Lambda_1] + \sin^2\phi \, R^\omega \Lambda_1 \Lambda_\omega \Lambda_1.
\end{aligned}
$$

Expanding the sums $R^\omega \Lambda_\omega$, one obtains

$$
\begin{aligned}
H'_a &= \cos^2\phi \, \left(R^0\Lambda_0 + R^1\Lambda_1 + R^2\Lambda_2 + R^3\Lambda_3\right) \\
&+ i\cos\phi\sin\phi \, R^2 \, [\Lambda_2, \Lambda_1] + i\cos\phi\sin\phi \, R^3 \, [\Lambda_3, \Lambda_1] \\
&+ \sin^2\phi \, \left(R^0\Lambda_1\Lambda_0\Lambda_1 + R^1\Lambda_1\Lambda_1\Lambda_1 + R^2\Lambda_1\Lambda_2\Lambda_1 + R^3\Lambda_1\Lambda_3\Lambda_1\right).
\end{aligned}
$$

By the lemmas 71 and 72, this simplifies to

$$
\begin{aligned}
H'_a &= \cos^2\phi \, \left(R^0\Lambda_0 + R^1\Lambda_1 + R^2\Lambda_2 + R^3\Lambda_3\right) \\
&+ 2\cos\phi\sin\phi \, R^2\Lambda_3 - 2\cos\phi\sin\phi \, R^3\Lambda_2 \\
&+ \sin^2\phi \, \left(R^0\Lambda_0 + R^1\Lambda_1 - R^2\Lambda_2 - R^3\Lambda_3\right).
\end{aligned}
$$

Hence:

$$
\begin{aligned}
H'_a &= R^0\Lambda_0 + R^1\Lambda_1 + \cos 2\phi \, R^2\Lambda_2 + \cos 2\phi \, R^3\Lambda_3 \\
&+ \sin 2\phi \, R^2\Lambda_3 - \sin 2\phi \, R^3\Lambda_2.
\end{aligned}
$$

The differential part H_d' is much simpler because $(\mathcal{D}\phi)$, which is an R-type quantion, commutes with every matrix Λ_i, which is an L-type quantion. Thus:

$$
\begin{aligned}
H_d' &= i\,(\cos\phi\, I - i\sin\phi\,\Lambda_1)\,(-\sin\phi\, I + i\cos\phi\,\Lambda_1)\,(\mathcal{D}\phi) \\
&= -\Lambda_1\,(\mathcal{D}\phi)\,.
\end{aligned}
$$

Collected into a vector, as in the expression (18.16), these results assume the form

$$
\begin{pmatrix} R^0 \\ R^1 \\ R^2 \\ R^3 \end{pmatrix}
\mapsto
\begin{pmatrix} R'^0 \\ R'^1 \\ R'^2 \\ R'^3 \end{pmatrix}
=
\begin{pmatrix} R^0 \\ R^1 - \mathcal{D}\phi \\ \cos 2\phi\; R^2 - \sin 2\phi\; R^3 \\ \sin 2\phi\; R^2 + \cos 2\phi\; R^3 \end{pmatrix}\,.
\tag{18.18}
$$

The group $U_2\,(1)$:

The transformation is

$$
E = \cos\phi\, I + i\sin\phi\,\Lambda_2 \in U_2\,(1)\,.
$$

For M', the same arguments as above yield

$$
M' = e^{-i2\phi\,\Lambda_2}\gamma^1\,.
\tag{18.19}
$$

Following the procedure written out in detail for the group $U_1\,(1)$, one obtains the following transformation relations for H :

$$
\begin{pmatrix} R^0 \\ R^1 \\ R^2 \\ R^3 \end{pmatrix}
\mapsto
\begin{pmatrix} R'^0 \\ R'^1 \\ R'^2 \\ R'^3 \end{pmatrix}
=
\begin{pmatrix} R^0 \\ \cos 2\phi\; R^1 + \sin 2\phi\; R^3 \\ R^2 - \mathcal{D}\phi \\ -\sin 2\phi\; R^1 + \cos 2\phi\; R^3 \end{pmatrix}\,.
\tag{18.20}
$$

The group $U_3\,(1)$:

The transformation is

$$
E = \cos\phi\, I + i\sin\phi\,\Lambda_3 \in U_3\,(1)\,.
$$

For M', Lemma 70 yields

$$
M' = E^\dagger \gamma^1 E^* = E^\dagger E \gamma^1\,.
$$

Hence,

$$M' = \gamma^1, \tag{18.21}$$

which is very different from the previous three cases.

By the same procedure as for U_1 one obtains the following transformations for H :

$$\begin{pmatrix} R^0 \\ R^1 \\ R^2 \\ R^3 \end{pmatrix} \mapsto \begin{pmatrix} R'^0 \\ R'^1 \\ R'^2 \\ R'^3 \end{pmatrix} = \begin{pmatrix} R^0 \\ \cos 2\phi\ R^1 - \sin 2\phi\ R^2 \\ \sin 2\phi\ R^1 + \cos 2\phi\ R^2 \\ R^3 - \mathcal{D}\phi \end{pmatrix}. \tag{18.22}$$

This completes the calculation of the effects of each of the four Abelian gauge groups $U_w\,(1)$ on the mass matrix and the matrix potential.

The infinitesimal transformations

For infinitesimal parameters ϕ^w, which will be written as $\varepsilon\phi^w$, so that the functions $\phi^w\,(x)$ would remain finite, the transformations derived above can be conveniently tabulated by listing only the differences, which are defined as

$$R^w \mapsto R^w + \delta R^w,$$
$$\gamma^1 \mapsto \gamma^1 + \delta M,$$

to first order in ε. Note that each entry in the following table has to be multiplied by ε.

	$U_0\,(1)$	$U_1\,(1)$	$U_2\,(1)$	$U_3\,(1)$
δR^0	$-\mathcal{D}\phi$	0	0	0
δR^1	0	$-\mathcal{D}\phi$	$2\phi\ R^3$	$-2\phi\ R^2$
δR^2	0	$-2\phi\ R^3$	$-\mathcal{D}\phi$	$2\phi\ R^1$
δR^3	0	$2\phi\ R^2$	$-2\phi\ R^1$	$-\mathcal{D}\phi$
δM	$-i2\phi\gamma^1$	$-i2\phi\gamma^6$	$-2\phi\gamma^3$	0

$$(18.23)$$

In the last row, $\gamma^6 = \Lambda_1\gamma^1$ and $\gamma^3 = \Lambda_2\gamma^1$.

18.2   Analysis of the results

Let us introduce some convenient terminology:

— The four R-type quantions R^0 to R^3 (or simply R when the distinction is not necessary) will be called **elementary potentials.**

— The four Abelian gauge groups $U_0(1)$ to $U_3(1)$ will be called **elementary gauge groups.**

— For every label ω, the elementary gauge group $U_\omega(1)$ and the elementary potential R^ω will be said to be **associated** to each other.

— At each point x, the ϕ^ω are the additive **parameters** of the four elementary gauge groups. As spacetime functions, $\phi^\omega = \phi^\omega(x)$, they will be referred to as **phase functions.**

We shall drop the label ω whenever the context allows no confusions.

Table (18.23) suggests several general observations:

(1) The differential gauge connections $(\mathcal{D}\phi)$ appear only on the diagonal, which means that the decomposition of the quantionic gauge group $U_q(1)$ into the mutually isomorphic Abelian groups $U_0(1)$ to $U_3(1)$ structurally matches the decomposition $H = R^\omega \Lambda_\omega$ of the matrix potential.

Since, for every ω, the parameter ϕ is additive and the operator $\mathcal{D}$ linear, the connections $(\mathcal{D}\phi)$ are the same for infinitesimal and finite gauge transformations.

(2) The elementary potential R^0 and its associated elementary group $U_0(1)$ are 'tightly paired', in the sense that the group $U_0(1)$ affects only the potential R^0, and that this potential is not affected by any other group.

(3) The algebraic part of the transformations of H has a simple geometric interpretation: At every spacetime point x, the elementary groups $U_1(1)$, $U_2(1)$, and $U_3(1)$ are formal rotations in a three-dimensional space spanned by the elementary potentials R^1, R^2, and R^3. The angle of rotation is $2\phi(x)$.

(4) For $\omega = 0, 1, 2$, the mass matrix acquires a phase factor. The global transformation is:

$$\gamma^1 \mapsto \exp\left(-i2\phi(x)\ \Lambda_\omega\right)\gamma^1.$$

Thus, computing the expressions $\Lambda_\omega \gamma^1$, one obtains

$$\left.\begin{array}{l}
\text{For } U_0: \ \gamma^1 \mapsto (\cos 2\phi - i \sin 2\phi) \, \gamma^1, \\
\text{For } U_1: \ \gamma^1 \mapsto \cos 2\phi \ \gamma^1 - i \sin 2\phi \ \gamma^6, \\
\text{For } U_2: \ \gamma^1 \mapsto \cos 2\phi \ \gamma^1 - \sin 2\phi \ \gamma^3, \\
\text{For } U_3: \ \gamma^1 \mapsto \gamma^1.
\end{array}\right\} \qquad (18.24)$$

We see that, in the case of $U_0 \, (1)$, the transformation (18.24) can be attributed to the mass parameter,

$$m \mapsto (\cos 2\phi - i \sin 2\phi) \, m,$$

while for $U_1 \, (1)$ and $U_2 \, (1)$ it necessarily modifies the matrix itself.

(5) The case $\omega = 3$ is distinguished by an invariant mass matrix.

The integrability condition

We see from the expressions (18.16), (18.18), (18.20), and (18.22) that every elementary gauge group $U_\omega \, (1)$ affects its associated elementary potential R only additively by way of the gauge connection $(-\mathcal{D}\phi)$:

$$R \mapsto R' = R - \mathcal{D}\phi. \qquad (18.25)$$

Thus, an elementary gauge connection is an elementary potential, but the opposite is not true.

Given an arbitrary R-type quantion R, the question arises whether it is a gauge connection. If it is, it is not viewed as a physical object because it can be eliminated by a gauge transformation; if it is not, the difference has the standard physical interpretation of a external potential.

In the familiar example of electromagnetism, the vector potential A_μ is a 'pure' connection if it is a gradient,

$$A_\mu = \partial_\mu \phi. \qquad (18.26)$$

For functions $A_\mu \, (x)$ of differentiability class at least C^2, the criterion for A_μ to be a gradient (18.26) is the integrability condition

$$F_{\mu\nu} \overset{def}{=} \partial_\nu A_\mu - \partial_\mu A_\nu = 0. \qquad (18.27)$$

We are to derive the corresponding criterion for R, where the condition corresponding to (18.26) is

$$R = (\mathcal{D}\phi). \qquad (18.28)$$

The simplest approach consists in expanding R and $\mathcal{D}$ in the pi basis defined in Section 3.5, where

$$\left. \begin{array}{l} \mathcal{D} = \Pi^\mu \partial_\mu, \\ R = \Pi^\mu A_\mu, \end{array} \right\} \qquad (18.29)$$

for an arbitrary vector field A_μ. In this formalism the condition (18.28) immediately assumes the form (18.26), and the criterion we are seeking is (18.27).

As in electromagnetism, the extent to which an elementary potential is not a gauge connection is characterized by an antisymmetric tensor $F_{\mu\nu}$ defined by the relation (18.27). The integrability condition $F_{\mu\nu} = 0$ characterizes a free quantionic field $|q\rangle$.

The structurally distinguished field equation

The elementary gauge group $U_3(1)$ is structurally distinguished as the only one that leaves invariant the right-hand side of the quantionic field equation (18.1). By limiting the matrix potential H to the associated elementary potential $R\Lambda_3$, one obtains the field equation

$$[\mathcal{D} - iR\Lambda_3]\,|q\rangle = im\gamma^1\,|q^*\rangle, \qquad (18.30)$$

which is form-invariant under the global group of elementary gauge transformations

$$E = (\cos\phi\, I + i\sin\phi\, \Lambda_3) \in U_3(1). \qquad (18.31)$$

Equation (18.30) will be referred to as the **quantionic Dirac equation.**

To obtain a better intuitive insight into this equation, we shall now prove its covariance in the matrix formalism. This will also help us to prepare the ground for the proof, in the next chapter, that quantions and Dirac's four-spinors are complementary formulations of the same object.

Beginning with the matrix E defined by the expression (18.31), we have

$$E = \begin{pmatrix} e^{i\phi} & 0 & 0 & 0 \\ 0 & e^{-i\phi} & 0 & 0 \\ 0 & 0 & e^{i\phi} & 0 \\ 0 & 0 & 0 & e^{-i\phi} \end{pmatrix}. \tag{18.32}$$

The mapping

$$|q) \longmapsto E\,|q)$$

yields the new equation

$$[\mathcal{D} - iR\Lambda_3]\,E\,|q) = im\gamma^1 E^*\,|q^*). \tag{18.33}$$

The relation

$$\gamma^1 E_1^* = E_1 \gamma^1$$

simplifies the right-hand side. To prove it, we compute both sides independently

$$\begin{aligned}
\gamma^1 E_1^* &= \begin{pmatrix} 0 & 0 & 0 & 1 \\ 0 & 0 & 1 & 0 \\ 0 & -1 & 0 & 0 \\ -1 & 0 & 0 & 0 \end{pmatrix} \begin{pmatrix} e^{-i\alpha} & 0 & 0 & 0 \\ 0 & e^{i\alpha} & 0 & 0 \\ 0 & 0 & e^{-i\alpha} & 0 \\ 0 & 0 & 0 & e^{i\alpha} \end{pmatrix} \\
&= \begin{pmatrix} 0 & 0 & 0 & e^{i\alpha} \\ 0 & 0 & e^{-i\alpha} & 0 \\ 0 & -e^{i\alpha} & 0 & 0 \\ -e^{-i\alpha} & 0 & 0 & 0 \end{pmatrix},
\end{aligned}$$

and

$$\begin{aligned}
E_1 \gamma^1 &= \begin{pmatrix} e^{i\alpha} & 0 & 0 & 0 \\ 0 & e^{-i\alpha} & 0 & 0 \\ 0 & 0 & e^{i\alpha} & 0 \\ 0 & 0 & 0 & e^{-i\alpha} \end{pmatrix} \begin{pmatrix} 0 & 0 & 0 & 1 \\ 0 & 0 & 1 & 0 \\ 0 & -1 & 0 & 0 \\ -1 & 0 & 0 & 0 \end{pmatrix} \\
&= \begin{pmatrix} 0 & 0 & 0 & e^{i\alpha} \\ 0 & 0 & e^{-i\alpha} & 0 \\ 0 & -e^{i\alpha} & 0 & 0 \\ -e^{-i\alpha} & 0 & 0 & 0 \end{pmatrix}.
\end{aligned}$$

The field equation (18.33) now reads

$$[\mathcal{D} - iR\Lambda_3] \, E \, |q) = imE\gamma^1 \, |q^*) \, .$$

Multiplying both sides from the left by $E^\dagger$ eliminates E from the right-hand side:

$$E^\dagger \, [\mathcal{D} - iR\Lambda_3] \, E \, |q) = im\gamma^1 \, |q^*) \, .$$

After expansion, we get

$$\left(E^\dagger \mathcal{D} E - iR E^\dagger \Lambda_3 E \right) |q) = im\gamma^1 \, |q^*) \, . \tag{18.34}$$

Since the matrix E, given by relation (18.31), is a linear combination of I and Λ_3, it commutes with Λ_3. this implies

$$E^\dagger \Lambda_3 E = \Lambda_3 \, .$$

Equation (18.34) now reads

$$E^\dagger \mathcal{D} E \, |q) - iR\Lambda_3 \, |q) = im\gamma^1 \, |q^*) \, . \tag{18.35}$$

The next step is to compute $\mathcal{D} E \, |q)$. By the Leibniz identity, we have

$$\begin{aligned} E^\dagger \mathcal{D} E \, |q) \;&=\; E^\dagger \left[(\mathcal{D} E) \, |q) + E\mathcal{D} \, |q) \right] \\ &=\; \left(E^\dagger \mathcal{D} E \right) |q) + \mathcal{D} \, |q) \, , \end{aligned}$$

so that Equation (18.35) becomes

$$\mathcal{D} \, |q) + \left[\left(E^\dagger \mathcal{D} E \right) - iR\Lambda_3 \right] |q) = im\gamma^1 \, |q^*) \, . \tag{18.36}$$

The only remaining term is $\left(E^\dagger \mathcal{D} E \right)$.

For $(\mathcal{D}E)$, direct computation yields

$$(\mathcal{D}E) = \begin{pmatrix} D & 0 & \delta & 0 \\ 0 & D & 0 & \delta \\ \delta^* & 0 & \Delta & 0 \\ 0 & \delta^* & 0 & \Delta \end{pmatrix} \begin{pmatrix} e^{i\alpha} & 0 & 0 & 0 \\ 0 & e^{-i\alpha} & 0 & 0 \\ 0 & 0 & e^{i\alpha} & 0 \\ 0 & 0 & 0 & e^{-i\alpha} \end{pmatrix}$$

$$= \begin{pmatrix} De^{i\alpha} & 0 & \delta e^{i\alpha} & 0 \\ 0 & De^{-i\alpha} & 0 & \delta e^{-i\alpha} \\ \delta^* e^{i\alpha} & 0 & \Delta e^{i\alpha} & 0 \\ 0 & \delta^* e^{-i\alpha} & 0 & \Delta e^{-i\alpha} \end{pmatrix}$$

$$= i \begin{pmatrix} e^{i\alpha} D\alpha & 0 & e^{i\alpha}\delta\alpha & 0 \\ 0 & -e^{-i\alpha} D\alpha & 0 & -e^{-i\alpha}\delta\alpha \\ e^{i\alpha}\delta^*\alpha & 0 & e^{i\alpha}\Delta\alpha & 0 \\ 0 & -e^{-i\alpha}\delta^*\alpha & 0 & -e^{-i\alpha}\Delta\alpha \end{pmatrix}$$

$$= i \begin{pmatrix} e^{i\alpha} & 0 & 0 & 0 \\ 0 & -e^{-i\alpha} & 0 & 0 \\ 0 & 0 & e^{i\alpha} & 0 \\ 0 & 0 & 0 & -e^{-i\alpha} \end{pmatrix} \begin{pmatrix} D\alpha & 0 & \delta\alpha & 0 \\ 0 & D\alpha & 0 & \delta\alpha \\ \delta^*\alpha & 0 & \Delta\alpha & 0 \\ 0 & \delta^*\alpha & 0 & \Delta\alpha \end{pmatrix}$$

$$= iE\Lambda_3 (\mathcal{D}\alpha).$$

Hence:

$$\left(E^\dagger \mathcal{D}E \right) = i\Lambda_3 (\mathcal{D}\alpha).$$

Finally, the transformed equation (18.30) reads, as expected,

$$[\mathcal{D} - i (R - \mathcal{D}\alpha) \Lambda_3] |q\rangle = im\gamma^1 |q^*\rangle. \tag{18.37}$$

The importance of the quantionic field equation (18.30) and of the gauge transformation (18.31) stems from their being structurally distinguished. As a reminder, a quantionic object is "structurally distinguished", or "inherent", if it emerge exclusively out of the theory of quantions augmented with Zovko's interpretation and structural quantization.

The physical interpretations of the equation (18.30) and of the group $U_3 (1)$ will be derived in the next chapter. To set the stage, let us preview the results.

(a) The quantionic Equation (18.30) is the quantionic form of Dirac's spinorial equation.

(b) The Hermitian R-type matrix R is the quantionic form of the electromagnetic relativistic potential A^μ.

(c) The quantionic gauge group $U_3(1)$ is the quantionic form of the complex gauge group of transformations of Dirac spinors.

(d) The quantionic derivation operator $\mathcal{D}$ is the quantionic form of Dirac's derivation operator $\gamma^\mu \partial_\mu$.

(e) The quantionic gauge connection $(\mathcal{D}\phi)$, by which the operator $\mathcal{D}$ is covariant, is the quantionic form of the complex gauge connection ieA^μ (where $\partial_\mu A^\mu = 0$), by which the operator $\gamma^\mu \partial_\mu$ is made covariant.

The standard coupling constants (like the charge e) have not been introduced in the present work because we have been concerned exclusively with the derivation of mathematical theorems. The role of the occasional excursions into physical interpretations is essentially to verify that the quantionic approach is not leading us away from the physics which is currently beyond doubt for being supported by experiments.

The 'non-distinguished' equations

The other three field equations are not form-invariant under the groups $U_0(1)$, $U_1(1)$, $U_2(1)$ because the mass matrix is not invariant but transforms according to the relations (18.24).

It is tempting to conjecture that this is why these equations, unlike Dirac's equation, are not semi-classical. We expect them to be meaningful after a 'second quantization', whose quantionic counterpart remains to be discovered. The table (18.23) supports this conjecture by suggesting the following identifications:

The photon: $H = R\Lambda_3$ (Proved in the next chapter.)

The neutral intermediate vector boson: $H = R\Lambda_0$.

The charged intermediate vector bosons: Some linear combinations of $H = R\Lambda_1$ and $H = R\Lambda_2$.

Chapter 19

The Quantion-Spinor Complementarity

The quantionic field equation (18.30) is distinguished as the only differential equation which is invariant under a nontrivial group of gauge transformations, that group being $U_3(1)$. The differential operator $[\mathcal{D} - iR\Lambda_3]$ in this equation is a covariant operator, meaning that the local gauge transformations $E \in U_3(1)$ affects it only by way of a differential connection which may be absorbed by the matrix potential R, thus making the equation form-invariant.

The question arises whether Equation (18.30) is already known in physics.

We have already encountered such a question in Chapter 15, where the equation to be interpreted was (15.13). It was immediately evident by the structure of this equation that the equivalent physical equation is Schrödinger's.

For Equation (18.30), however, no immediate interpretation is possible by comparison with some well-known physical equation because the quantionic formalism is not standard. We thus have to look for reformulations of this equation in some formalism that might be recognizable.

Such a reformulation exists and it is unique. We shall develop it in the next section.

19.1   The spinorial Dirac equation

The first objective is to find all formulations that are equivalent in content to Equation (18.30), but differ from it in form. To this end, we consider this equation together with its complex conjugate:

$$[\mathcal{D} - iR\Lambda_3]\,|q) = im\gamma^1\,|q^*)\,, \tag{19.1}$$
$$[\mathcal{D}^* + iR^*\Lambda_3]\,|q^*) = -im\gamma^1\,|q)\,. \tag{19.2}$$

For the most general matrix $R \in \mathcal{R}^h$, we write

$$R = \begin{pmatrix} r & 0 & c & 0 \\ 0 & r & 0 & c \\ c^* & 0 & s & 0 \\ 0 & c^* & 0 & s \end{pmatrix}, \tag{19.3}$$

where the spacetime functions r and s are real and the function c is complex. Then:

$$\mathcal{D} - iR\Lambda_3 = \begin{pmatrix} D - ir & 0 & \delta - ic & 0 \\ 0 & D + ir & 0 & \delta + ic \\ \delta^* - ic^* & 0 & \Delta - is & 0 \\ 0 & \delta^* + ic^* & 0 & \Delta + is \end{pmatrix}. \tag{19.4}$$

Expanding Equation (19.1), one obtains four complex equations:

$$\begin{aligned} (D - ir)\,q_1 + (\delta - ic)\,q_3 &= imq_4^*, \\ (D + ir)\,q_2 + (\delta + ic)\,q_4 &= imq_3^*, \\ (\delta^* - ic^*)\,q_1 + (\Delta - is)\,q_3 &= -imq_2^*, \\ (\delta^* + ic^*)\,q_2 + (\Delta + is)\,q_4 &= -imq_1^*. \end{aligned}$$

The expansion of Equation (19.2) yields the complex conjugates of these four equations:

$$\begin{aligned} (D + ir)\,q_1^* + (\delta^* + ic^*)\,q_3^* &= -imq_4, \\ (D - ir)\,q_2^* + (\delta^* - ic^*)\,q_4^* &= -imq_3, \\ (\delta + ic)\,q_1^* + (\Delta + is)\,q_3^* &= imq_2, \\ (\delta - ic)\,q_2^* + (\Delta - is)\,q_4^* &= imq_1. \end{aligned}$$

We now have a system of eight equations, out of which only four are mutually independent.

In these sets, the four unknown functions q_1 to q_4 and their complex conjugates appear on different sides of the equations. All eight functions are thus mutually coupled.

The question is whether these eight equations can be restructured as two decoupled sets of four equations. They can, and the solution is essentially unique.

If we take q_1, q_2^*, q_3, q_4^* to be the unknowns, the two sets of equations acquire a different structure. One set of four equations is

$$
\begin{aligned}
(D - ir)\, q_1 + (\delta - ic)\, q_3 &= imq_4^*, \\
(\delta^* - ic^*)\, q_1 + (\Delta - is)\, q_3 &= -imq_2^*, \\
(D - ir)\, q_2^* + (\delta^* - ic^*)\, q_4^* &= -imq_3, \\
(\delta - ic)\, q_2^* + (\Delta - is)\, q_4^* &= imq_1.
\end{aligned}
$$

The second set is

$$
\begin{aligned}
(D + ir)\, q_2 + (\delta + ic)\, q_4 &= imq_3^*, \\
(\delta^* + ic^*)\, q_2 + (\Delta + is)\, q_4 &= -imq_1^*, \\
(D + ir)\, q_1^* + (\delta^* + ic^*)\, q_3^* &= -imq_4, \\
(\delta + ic)\, q_1^* + (\Delta + is)\, q_3^* &= imq_2.
\end{aligned}
$$

These sets of equations are mutually related by complex conjugation but not differentially, so that one set may be ignored as redundant.

Taking the first set, let us re-arrange it as follows:

$$
\begin{aligned}
(D - ir)\, q_2^* + (\delta^* - ic^*)\, q_4^* &= -imq_3, \\
(\delta - ic)\, q_2^* + (\Delta - is)\, q_4^* &= -im\,(-q_1), \\
-(\delta^* - ic^*)\,(-q_1) + (\Delta - is)\, q_3 &= -imq_2^*, \\
(D - ir)\,(-q_1) - (\delta - ic)\, q_3 &= -imq_4^*.
\end{aligned}
$$

This may be written as a single matrix equation

$$
\left(\tilde{D} - iA\right)
\begin{pmatrix}
q_3 \\
-q_1 \\
q_2^* \\
q_4^*
\end{pmatrix}
= -im
\begin{pmatrix}
q_3 \\
-q_1 \\
q_2^* \\
q_4^*
\end{pmatrix}.
\tag{19.5}
$$

The operator $\tilde{\mathcal{D}}$ is defined as

$$\tilde{\mathcal{D}} \stackrel{def}{=} \begin{pmatrix} 0 & 0 & D & \delta^* \\ 0 & 0 & \delta & \Delta \\ \Delta & -\delta^* & 0 & 0 \\ -\delta & D & 0 & 0 \end{pmatrix}, \tag{19.6}$$

and the matrix A as

$$A \stackrel{def}{=} \begin{pmatrix} 0 & 0 & r & c^* \\ 0 & 0 & c & s \\ s & -c^* & 0 & 0 \\ -c & r & 0 & 0 \end{pmatrix}.$$

Substitution of the definitions of Newman-Penrose symbols into the matrix $\tilde{\mathcal{D}}$ yields

$$\tilde{\mathcal{D}} = \begin{pmatrix} 0 & 0 & \partial_0 + \partial_3 & \partial_1 - i\partial_2 \\ 0 & 0 & \partial_1 + i\partial_2 & \partial_0 - \partial_3 \\ \partial_0 - \partial_3 & -\partial_1 + i\partial_2 & 0 & 0 \\ -\partial_1 - i\partial_2 & \partial_0 + \partial_3 & 0 & 0 \end{pmatrix}.$$

Thus, $\tilde{\mathcal{D}}$ is obviously the Dirac operator

$$\tilde{\mathcal{D}} = \gamma^\mu \partial_\mu. \tag{19.7}$$

Similarly, we may write the matrix A in terms of the four real components of a vector field $A_\mu(x)$:

$$A = \begin{pmatrix} 0 & 0 & A_0 + A_3 & A_1 - iA_2 \\ 0 & 0 & A_1 + iA_2 & A_0 - A_3 \\ A_0 - A_3 & -A_1 + iA_2 & 0 & 0 \\ -A_1 - iA_2 & A_0 + A_3 & 0 & 0 \end{pmatrix},$$

so that

$$A = \gamma^\mu A_\mu. \tag{19.8}$$

It follows from these results that the relationship between quantions and Dirac spinors Ψ is uniquely determined up to a factor:

$$\begin{pmatrix} q_3 \\ -q_1 \\ q_2^* \\ q_4^* \end{pmatrix} \equiv k \begin{pmatrix} \psi_1 \\ \psi_2 \\ \psi_3 \\ \psi_4 \end{pmatrix}. \tag{19.9}$$

At this point, k is an arbitrary real coefficient.

Equation (19.5) now assumes the well-known form

$$[\gamma^\mu (\partial_\mu - iA_\mu) + im]\, \Psi = 0. \tag{19.10}$$

This proves that the quantionic Dirac equation is equivalent to the standard spinorial version.

19.2   The Dirac current

As in Section 15.5 for the Schrödinger current, we shall verify in this section that the quantionic current coincides with the Dirac current.

We begin by taking the inverse of the correspondence (19.9):

$$|q) = \begin{pmatrix} q_1 \\ q_2 \\ q_3 \\ q_4 \end{pmatrix} \equiv k \begin{pmatrix} -\psi_2 \\ \psi_3^* \\ \psi_1 \\ \psi_4^* \end{pmatrix}. \tag{19.11}$$

By transforming the q-ket $|q)$ to the linked quantion Q, one obtains

$$Q = (\Omega|q) = k \begin{pmatrix} -\psi_2 & \psi_1 & 0 & 0 \\ \psi_3^* & \psi_4^* & 0 & 0 \\ 0 & 0 & -\psi_2 & \psi_1 \\ 0 & 0 & \psi_3^* & \psi_4^* \end{pmatrix}. \tag{19.12}$$

By Zovko's interpretation, the quantionic current $j = A(Q)$ is

$$
\begin{aligned}
j_{red} &= \left(Q^\dagger Q\right)_{red} = k^2 \begin{pmatrix} -\psi_2^* & \psi_3 \\ \psi_1^* & \psi_4 \end{pmatrix} \begin{pmatrix} -\psi_2 & \psi_1 \\ \psi_3^* & \psi_4^* \end{pmatrix} \\
&= k^2 \begin{pmatrix} \psi_2^*\psi_2 + \psi_3\psi_3^* & -\psi_2^*\psi_1 + \psi_3\psi_4^* \\ -\psi_1^*\psi_2 + \psi_4\psi_3^* & \psi_1^*\psi_1 + \psi_4\psi_4^* \end{pmatrix}.
\end{aligned}
$$

It follows from relations (2.29) that the vector j^μ is

$$j^\mu = \frac{k^2}{2} \begin{pmatrix} \psi_2^*\psi_2 + \psi_3\psi_3^* + \psi_1^*\psi_1 + \psi_4\psi_4^* \\ -\psi_2^*\psi_1 + \psi_3\psi_4^* - \psi_1^*\psi_2 + \psi_4\psi_3^* \\ i\,(-\psi_2^*\psi_1 + \psi_3\psi_4^* + \psi_1^*\psi_2 - \psi_4\psi_3^*) \\ \psi_2^*\psi_2 + \psi_3\psi_3^* - \psi_1^*\psi_1 - \psi_4\psi_4^* \end{pmatrix}. \tag{19.13}$$

This specifies k :

$$k = \sqrt{2}. \tag{19.14}$$

The expression (19.13) for the Minkowski current is formally unsatisfactory for being written in terms of the components of the spinor Ψ. Since the structural object is Ψ, and since the current is real, the only structurally admissible expression is a Hermitian scalar product of the form

$$j^\mu = \Psi^\dagger \Gamma^\mu \Psi, \tag{19.15}$$

where the matrices Γ^μ of numerical coefficients are Hermitian.

Comparison of the expression (19.15), that is

$$j^\mu = \begin{pmatrix} \psi_1^* & \psi_2^* & \psi_3^* & \psi_4^* \end{pmatrix} \Gamma^\mu \begin{pmatrix} \psi_1 \\ \psi_2 \\ \psi_3 \\ \psi_4 \end{pmatrix},$$

with the relations (19.13) yields the matrices Γ^μ :

$$\Gamma^0 \overset{def}{=} I,$$

$$\Gamma^1 \overset{def}{=} \begin{pmatrix} 0 & -1 & 0 & 0 \\ -1 & 0 & 0 & 0 \\ 0 & 0 & 0 & 1 \\ 0 & 0 & 1 & 0 \end{pmatrix} = \gamma^0 \gamma^1,$$

$$\Gamma^2 \overset{def}{=} \begin{pmatrix} 0 & i & 0 & 0 \\ -i & 0 & 0 & 0 \\ 0 & 0 & 0 & -i \\ 0 & 0 & i & 0 \end{pmatrix} = \gamma^0 \gamma^2,$$

$$\Gamma^3 \overset{def}{=} \begin{pmatrix} -1 & 0 & 0 & 0 \\ 0 & 1 & 0 & 0 \\ 0 & 0 & 1 & 0 \\ 0 & 0 & 0 & -1 \end{pmatrix} = \gamma^0 \gamma^3.$$

The quantionic current in spinorial form is thus

$$j^\mu = \Psi^\dagger \gamma^0 \gamma^\mu \Psi, \tag{19.16}$$

which is the standard expression for the Dirac current.

The coincidence of the quantionic and spinorial forms of the Dirac current is not unexpected (except for the numerical value of k), but its detailed proof plays a useful role in confirming the integrity of the calculations that lead to it.

19.3 Discussion

At this point, the following question suggests itself:

Since the quantionic and the spinorial approach yield isomorphic forms of the Dirac equation, what are the advantages of the quantionic approach?

Several observations answer this question.

The inherence conjecture

Let us visualize standard physics as a tree rooted in mathematics. In this picture, simplified to a one-dimensional bamboo, a node represents a system of basic concepts and postulates together with all mathematically derived concepts and theorems.

The mathematical foundations are the number systems (real and complex) and real analysis.

Let the first node be the postulate which bridges the two millennia that separate Newton from Aristotle. That postulate is the law of inertia, according to which a system has, at every instant t, a memory of its state of motion at the infinitesimally close instant in its past, $t - dt$. According to Aristotle, it has no memory.

The second node adds the postulate that a system's memory can be externally modified only in the simplest possible way, which is additively. This gives rise to classical dynamics, where motions are governed by second order differential equations in time. These equations would be of first order if Aristotle had his way, and of third or higher order in a less obliging universe.

Granularity of action is introduced at the third node. This gives rise to the Schrödinger equation, which is still of second order in time.

The fourth node introduces the relativistic structure of spacetime. The Schrödinger equation gives way to the Dirac equation, which is of first order in time — but at the cost of additional degrees of freedom.

Thus, Dirac's equation is on the fourth level above ground, and each level was a major discovery in physics — it's being understood that a physical discovery manifests itself as a new postulate.

In contrast, the quantionic approach is streamlined:

First: Adjoin the theory of quantions to the mathematical foundations.

Second: Define the first level above ground as all mathematical theorems that follow from Zovko's interpretation of the algebra of quantions. All of physics so derived — referred to as "inherent" — is at this level.

Since the quantionic version of Dirac's equation resides at this first level, it is both inherent and contains more physics than the equation derived by Dirac at the fourth level of the standard approach — even though the two are identical in the spinorial formalism. (If they were not identical, the quantionic version would clash with observations and this book would not have been written.)

By the still 'privately held' conjecture of inherence — which is the author's guiding idea in the investigation of the quantionic approach — all of known physics would be inherent. The difficulty is in discovering the relevant quantionic concepts and theorems, as many of them differ too much from the corresponding standard concepts to be evident.

A path to electroweak concepts

The quantionic approach is equivalent to the spinorial approach only if one restricts the potential matrix H to the special case $R\Lambda_3$ — which case represents an electromagnetic potential. But it is evident from Table (18.23) that the elementary potential $R\Lambda_3$ is closely related to the potential R^0, and that the quantionic Dirac equation is valid for the more general potential

$$H = R^0 + R^3 \Lambda_3$$

if one admits a phase transformation

$$m \mapsto e^{-i2\chi}m$$

of the mass parameter.

The corresponding elementary gauge groups are $U_0(1)$ and $U_3(1)$. This suggests an analogy with the pairing of the photon with the neutral vector boson Z_0. What's more, the other two potentials $(R^1$ and $R^2)$ in the complete decomposition

$$H = R^0 + R^1\Lambda_1 + R^2\Lambda_2 + R^3\Lambda_3$$

are also mutually related in a manner that vaguely suggests the relationship between the intermediate vector bosons W_1 and W_2.

Setting aside the mass matrix, which transforms in a very unpleasant way, the gauge group of the full matrix potential H is the quantionic gauge group $U_q(1)$. This group in mathematically distinguished because it plays, in the algebra of quantions, the role played by the gauge group $U(1)$ in the field of complex numbers. But this group is also physically distinguished for being isomorphic with the gauge group $U(1) \times SU(2)$ of the electroweak unification.

We have here such a confluence of coincidences that the expected interpretations seem unavoidable. Yet, at this point, an exact comparison cannot be made between electroweak and quantionic concepts because the latter are still semi-classical.

A final observation

The structural unification developed so far is based on the idea that quantions are simultaneously algebraic and relativistic. Yet, the fields $|q\rangle$ are manifestly neither. To exhibit their algebraic structure, one must invoke the concept of linking; to exhibit their geometric spacetime structure, one must invoke the concept of complementarity.

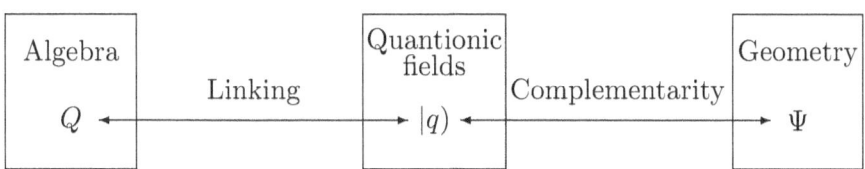

Annotated Bibliography on Quantions

[1] Grgin, Emile and Petersen, Aage. Algebraic Implications of Composability of Physical Systems. Commun. Math. Phys. 50 177-188 (1976).

This was the first attempt at developing an abstract view of quantum mechanics in the expectation that it might lead to a better understanding of the incompatibility of this theory with relativity — though Petersen, more sanguine than myself, had less modest expectations. The article contains two topics: an abstract characterization of quantum mechanics, and the algebraic formulation of this theory in phase space. It is the abstract formulation of quantum mechanics which led to quantions many years later. The long hiatus was due in part to our having overlooked an essential idea without which continuation was virtually impossible (see under [6]). Mostly out of curiosity, the second topic was revisited in the following two-part paper.

[2] Grgin, Emile and Sandri, Guido. "The Quantum Oscillator in Phase Space" Part I, Fizika B 5 2, 141-158 (1996). (Online at http://fizika.hfd.hr)

[3] Grgin and Sandri. "The Quantum Oscillator in Phase Space" Part II, Fizika B 6 1, 1-22 (1997). (http://fizika.hfd.hr).

In this work, the phase space formulation of quantum mechanics developed in [1] is tested on the quantum mechanical harmonic oscillator. The solution agrees completely with the Hilbert space solution. It also provides additional insights into the relationship between quantum and classical mechanics. This encouraging result motivated the present author to return, twenty years later, to the work initiated with Petersen, but that had been dropped after running into a snag.

[4] Grgin, Emile. "Relativistic ring extension of the field of complex numbers". Physics Letters B 431, 15-18 (1998).
This short paper contains the first indication of the existence of a relativistic algebraic extension of the field of complex numbers.

[5] Grgin, Emile. "Quantions and Inherently Relativistic Quantum Theory". In Festschrift in Honor of Paolo Budinich. Bradamante and Furlan, Editors. Consorzio per l'Incremento degli Studi e delle Ricerche dei Dipartimenti di Fisica dell'Università di Trieste. 94119 (2001).
This is the contents of a lecture at a multidisciplinary gathering in Lošinj (Croatia) on some aspects of the work, then in progress, which was subsequently published in the papers [6] to [9].

[6] Grgin, Emile. "Inherently Relativistic Quantum Theory. Part I. The Algebra of Observables". Fizika, B (Zagreb) 10 113-138 (2001).
(On line at http://fizika.hfd.hr)
This paper supersedes and completes the abstract approach to quantum mechanics initiated in [1] in collaboration with Peterson. It concludes with an abstract formulation of quantum mechanics as a Lie algebra which is also a Jordan algebra, and in which two other identities hold in addition to the Jacobi identity. This result differs from the result obtained with Petersen in that the latter was formulated too narrowly. As a consequence, it missed the possibility that $\sqrt{-I}$ (where I is some unit matrix) could have real non-Hermitian solutions. This is what prevented Petersen and myself from seeing how to continue on the abstract path. Once this flaw had been corrected in [6], the abstract approach became almost exclusively deductive. The first step in this direction was the following paper.

[7] Grgin, Emile. "Inherently Relativistic Quantum Theory. Part II. Classification of the Solutions". Fizika, B (Zagreb) 10 139-160 (2001). (On line at http://fizika.hfd.hr)

An extension of Cartan's classification is applied in this paper to the abstract structure developed in [6]. The outcome is that there are exactly two realization: The Lie algebras $su\,(n)$, that is, standard quantum mechanics in Hilbert space, and the Lie algebra $so\,(2,4)$. Since $SO(2,4)$ is the conformal group of Minkowski space, it follows that one realization is inherently relativistic.

[8] Grgin, Emile. "Inherently Relativistic Quantum Theory. Part III. Quantionic Algebra". Fizika, B (Zagreb) 10 195-218 (2001). (On
line at http://fizika.hfd.hr)
[9] Grgin, Emile. "Inherently Relativistic Quantum Theory. Part IV. Quantionic Theorems". Fizika, B (Zagreb) 10 219-242 (2001). (On line at http://fizika.hfd.hr)
In these two papers, the solution based on the algebra $so\,(2,4)$ is re-parametrized without superfluous variables. This yields the spinless version of the algebra of quantions.

[10] Grgin, Emile: *"The Algebra of quantions — A Unifying Number System for Quantum Mechanics and Relativity"*. AuthorHouse, Bloomington, Indiana, (2005). www.authorhouse.com.
Many chapters of this book are essentially essays on the abstract approach to quantum mechanics and on the applications of the algebra of quantions. The mathematical chapters are to be viewed as preprints of papers never submitted for publication. The main contribution is a derivation of the algebra of quantions as an extension of the field of complex numbers which is unique from a point of view that has never been considered in mathematics (the elimination of an overlooked degeneracy of the complex numbers). This version of the algebra of quantions contains spin. It is the version postulated in the present volume. This book also contains several potentially interesting ideas that remain to be carefully investigated.

For readers who might be interested in the abstract origins of the algebra of quantions, the relevant papers are [6] to [9] and the first five chapters of the book [10]. The other papers are either digressions (the case of [2] and [3]), or are being superseded by the present work, which is self-contained.

Index

.

www.ingramcontent.com/pod-product-compliance
Lightning Source LLC
Chambersburg PA
CBHW031831170526
45157CB00001B/256